Principles of Poultry Science

S.P. Rose
Senior Lecturer
Harper Adams Agricultural College
Newport
Shropshire TF10 8NB
UK

D0145739

CAB INTERNATIONAL

CABI Publishing is a division of CAB International

CABI Publishing
CAB International
Wallingford
Oxon OX10 8DE
UK

Tel: +44 (0)1491 832111
Fax: +44 (0)1491 833508
Email: cabi@cabi.org
Web site: www.cabi-publishing.org

CABI Publishing
875 Massachusetts Avenue
7th Floor
Cambridge, MA 02139
USA

Tel: +1 617 395 4056
Fax: +1 617 354 6875
Email: cabi-nao@cabi.org

A catalogue record for this book is available from the British Library, London, UK.
A catalogue record for this book is available from the Library of Congress, New York, USA.

ISBN 0 85199 122 X

First printed 1997
Reprinted 2001, 2005

Printed and bound in the UK by Biddles Ltd, Kings Lynn

Contents

Preface

Poultry production is an increasingly important agricultural industry in the world. Poultry meat and eggs account for about 10% of the total weight of all meat, milk and eggs produced in the world each year. The poultry industry has a continuing need for technologists that understand the scientific principles relating to poultry production.

Few specialist courses in poultry science still exist. Poultry science, and its application in the poultry industry, are increasingly being taught as part of courses of animal science or agriculture within the universities and colleges. Students therefore need to understand a wide subject area. Teaching in any one subject area, for example poultry, must concentrate on only the most important scientific principles that affect the production system. Although many universities and colleges no longer hold large poultry flocks, poultry are still frequently used in research and investigational projects by students. These students often have an urgent need to understand some aspects of poultry science that relate to their project work.

The purpose of this book is to give undergraduate or postgraduate students a relatively brief description of the most important aspects of poultry science. Some suggestions of further reading are given for people who wish to have more detailed information in a particular subject area. The book deals with all the commercially important poultry species – domestic fowl, turkeys, quail, guinea fowl, ducks and geese – although there is an emphasis towards domestic fowl because of their importance in the poultry industry of most countries of the world. The information given is relevant to most types of production systems in a variety of climates. People using this book may have a greater understanding of other farm animal species, so the differences and similarities between poultry and other farm animals are stressed.

Poultry scientists frequently use mathematical modelling techniques to describe the responses of growing and egg laying poultry.

Modern generations of students are increasingly more competent at using computers to calculate relatively complicated mathematical models. The use of a computer spreadsheet programme can rapidly display graphs that describe bird responses predicted from mathematical models. The book emphasizes the use of these models and gives examples and the relevant coefficients. I hope that this information will allow students to use computers to adopt a more interesting and interactive approach to their understanding of the biological responses of poultry.

Domesticated Poultry: A Description

1

TAXONOMY

Poultry are birds that have been selected and domesticated by man. Two taxonomic families of birds have been the easiest to domesticate. Phasianidae are pheasant-type game birds that include chickens, turkeys, quail and guinea fowl. The Anatidae family are waterfowl that include ducks and geese (see Fig. 1.1).

More than a century of intense poultry breeding has resulted in a large diversity between the strains of some poultry species. Crossing closely related poultry species to produce hybrid offspring has the potential to increase the diversity of poultry even further. However, only one such cross has ever been used in commercial production systems. The cross of Muscovy ducks (*Cairina moschata*) with common ducks (*Anas platyrhynchos*) produces a viable but sterile hybrid. These offspring are often called mules and are used for duck meat production and fat liver production in some countries.

There is no commercial exploitation of inter-species hybrids within the Phasianidae. Some crosses, for example the chicken × turkey, produce fertile eggs that do not hatch. Other crosses, for example the chicken × guinea fowl, give offspring that may hatch but they die shortly afterwards.

DOMESTICATION AND IMPROVEMENT OF POULTRY

Poultry were domesticated later than other farm livestock. Sheep, pigs and cattle were probably first domesticated in southwest Asia in 9000BC, 7000BC and 6000BC respectively. Domesticated fowl from Asia in 2500BC are probably the main source of modern stocks. Domesticated geese were established in Egypt in 1500BC and the earliest evidence of turkey domestication was in Mexico in 2500BC. Ducks were domesticated in

China in 2500BC but there was no successful domestication in the West until around 1200AD. Although domesticated Muscovy ducks were found in Peru in the 16th century AD, it is likely the domestication had happened recently.

Sheep, pigs and cattle were, from the first stages of domestication, primarily used as a food source. Poultry were initially domesticated for religious, cultural and entertainment reasons and only later thought of as a food source. The placid nature of the duck to a large extent explains its late domestication in Europe compared to the chicken, which is an aggressive and spectacular fighter! Quail were first domesticated in Asia around the 11th century AD, but their main use was as caged song birds. Interest in their use for edible eggs and meat began in Japan only in the early 20th century.

Around 1850 there was an explosion of interest in domesticated poultry and large sums of money were spent acquiring, breeding and writing about these birds. During this period, called the 'hen craze',

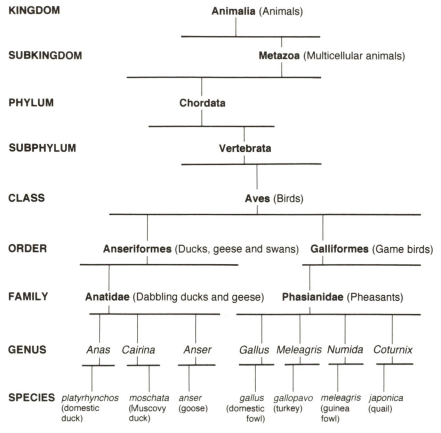

Fig. 1.1. A classification hierachy of domesticated poultry.

most of the breeds of chickens that now dominate world poultry production systems were developed. Many variously coloured and shaped breeds were identified and developed. The external anatomy of a domestic fowl is shown in Fig. 1.2.

The objectives of the early poultry breeders of domestic fowl were to achieve perfection in show specimens and their breed standards and they had little interest in the efficiency of productive performance. However, one breed, the single comb White Leghorn, was used by commercial breeders in the 1950s as a highly productive white egg layer. In the 1960s brown egg laying strains were derived by crossing Rhode Island Reds with White Leghorns and other minor breeds. Domestic fowl used for meat production were based on a cross of the Cornish with the Plymouth White Rock breeds. The numbers of companies breeding commercial poultry decreased rapidly after the 1950s. Each company must support a large breeding programme so must therefore rely on large, worldwide sales of their strains to recoup these costs. There are now fewer than a dozen large multinational companies that breed meat birds and a similar number of companies that breed egg laying stocks.

Early settlers to North America realized the potential value of domesticating turkeys. By the early 17th century they had crossed two turkey subspecies to produce a turkey with acceptable growth characteristics. This became known as the American Bronze and this strain was used with little further change until the 20th century. Turkey breeders

Fig. 1.2. The external anatomy of a domestic fowl.

in the 1920s selected for large birds with wide breasts. Within a few years most North American breeding stocks had changed from the narrow-breasted early types to the new broad-breasted turkeys. The increase in size and muscling was greatest in males. Mature male and female turkeys have very different mature body weights in modern strains. This has led to problems in their being unable to mate successfully. Artificial insemination techniques have had to be developed to allow most turkey breeding programmes to continue.

The USA government began a breeding programme in the 1950s to produce a small turkey with a good meat yield. This strain was called the Beltsville Small White and it was the forerunner of modern strains of small broiler turkeys.

Genetic selections of modern strains of guinea fowl and quail only began in the 1950s. Guinea fowl are highly suited to the use of artificial insemination. Rapid advances in their selection programmes have been made since this was developed in the 1960s.

Selection of domesticated ducks has increased body weight and resulted in a loss of the ability to fly in common ducks, but not in Muscovy ducks. Heavy duck breeds such as Pekins, Aylesburys and Rouens were initially used as meat birds and Khaki Campbells and Indian Runners were used as egg laying strains. Intense selection programmes for egg laying and meat strains were started by large commercial breeding companies in the 1960s.

The dominant breeds of geese have been established for many centuries. The Emden is the prominent meat bird in Europe and America and the Toulouse is commonly used for fat liver production.

THE POULTRY POPULATION AND ITS DISTRIBUTION

Domestic fowl, ducks and turkeys are the three most common species of domesticated poultry in the world. Domestic fowl account for over 90% of the total. There are few social or religious taboos concerned with eating poultry products so poultry are kept for food production in most areas of the world. World poultry numbers are shown in Fig. 1.3.

The populations of egg laying domestic fowl very approximately equal the human populations of most countries of the world. Although three-quarters of the world's population live in developing countries, they produce less than one-third of the world's output of poultry meat and eggs. Poultry in developing countries are mostly kept in extensive systems where they may have to forage for limited amounts of feed. The productive outputs of these birds may be much lower than can be achieved in intensive production systems.

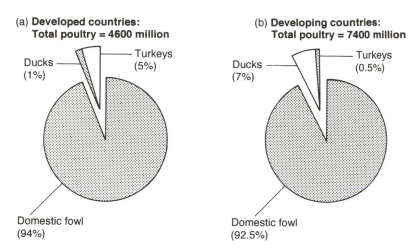

Fig. 1.3. World poultry numbers in 1992 (adapted from FAO, 1992).

Ducks are the next most common species of poultry in the world. There are large numbers in many countries in Asia. Ducks, like domestic fowl, are good scavengers and foragers. They can thrive in environments where feed supplies are limiting. Turkeys are more commonly reared in developed countries. Their potentials for fast growth rates and high meat yields are best exploited when nutrient dense feeds are given in intensive production systems.

The world production of poultry meat has increased throughout the 20th century. Intensive broiler chicken and turkey production methods were first developed in the 1950s. Poultry meat production was less than 10% of the world's total meat output at that time. Poultry meat output has taken a greater share of the expanding world meat market since then and it accounted for over 20% of total meat output by the mid 1980s (Fig. 1.4).

World egg production trebled between 1950 and 1985. The rate of increase of egg production in developed countries slowed in the 1970s but this was counterbalanced by a rapid expansion in egg production in developing countries (Fig. 1.5).

Social and economic factors have a large influence on changes in poultry numbers within a country. The changes in the poultry population of Great Britain are shown as an example in Fig. 1.6. The British poultry flock has increased seven fold in the last century, but the rate of increase has been erratic. This has been caused by a variety of historical events. Reductions in poultry numbers occurred around the 1st World War, the Depression of the 1930s and the 2nd World War. The rapid

Chapter 1

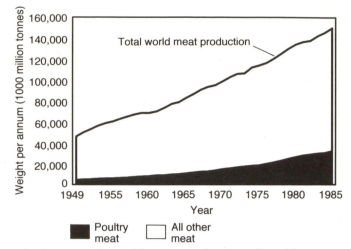

Fig. 1.4. Yearly changes in the world's meat production (adapted from FAO, 1987).

increase in poultry numbers after the 2nd World War was mainly due to two factors. First, there was a change from importing large numbers of eggs in 1950 towards being wholly self-sufficient in egg production by the 1970s. Second, the rapid development of the poultry meat industry in the 1950s and 1960s increased the numbers of broiler chickens, meat ducks and turkeys.

Poultry numbers started to decline in Britain in the 1970s. This coincided with a national trend for the human population to eat fewer eggs (see Fig. 1.7). Changes towards eating fewer cooked breakfasts and

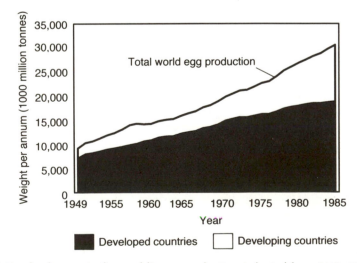

Fig. 1.5. Yearly changes in the world's egg production (adapted from FAO, 1987).

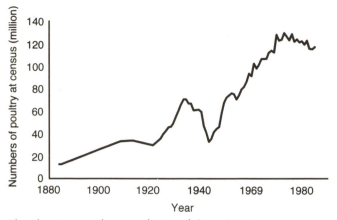

Fig. 1.6. The changing poultry population of Great Britain.

less fried food were the main causes of these eating patterns. The reduced demand for eggs has resulted in a smaller national laying hen flock.

Lower poultry meat prices led to increased consumption during the 1960s and 1970s. During the 1980s the increase in the sales of whole poultry carcasses slowed but there was an increase in the sales of further processed poultry meat products. The continued increased total demand for poultry meat has only resulted in a small increase in meat-bird numbers. There have been large improvements in growth rates during this time so that, although the numbers of poultry houses have not increased since the 1970s, the number of birds grown within a house each year has increased considerably.

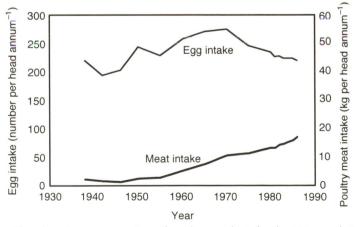

Fig. 1.7. The changing consumption of poultry products by the UK population (adapted from Marks, 1989).

FURTHER READING

Taxonomy

Delacour, J. (1964) *The Waterfowl of the World*. Volume 4. Country Life Limited, London.

Delacour, J. (1977) *The Pheasants of the World*. 2nd edn. Spur Publications, Hindhead, UK.

Stevens, L. (1991) *Genetics and Evolution of the Domestic Fowl*. Cambridge University Press, Cambridge.

Storer, R.W. (1960) The classification of birds. In: Marshall, A.J. (ed.) *Biology and Comparative Physiology of Birds*. Volume 1. Academic Press, New York, pp. 57–94.

Domestication

Crawford, R.D. (1990) Origin and history of poultry species. In: Crawford, R.D. (ed.) *Poultry Breeding and Genetics*. Elsevier, Amsterdam, pp. 1–42.

Schorger, A.W. (1966) *The Wild Turkey. Its History and Domestication*. University of Oklahoma Press, Norman.

West, B. and Zhou, B. (1989) Did chickens go north? New evidence for domestication. *World's Poultry Science Journal* 45, 205–218.

The Poultry Population and its Distribution

FAO (1987) *1948–1985 World Crop and Livestock Statistics*. FAO, Rome.

FAO (1992) *Production*. FAO Statistics Series No. 112, FAO, Rome.

Marks, H.F. (1989) *A Hundred Years of British Food and Farming. A Statistical Survey*. Taylor and Francis, London.

The Products   2

Poultry convert the nutrients in low value feedstuffs into high value meat and eggs. Both these poultry products are highly palatable, easily digestible and easily marketed. This chapter describes the basic characteristics and yields of poultry meat and eggs. It also explains the factors that affect the quality of these products.

POULTRY CARCASSES

Carcass Yields

A bird is usually bled and then plucked directly after it has been slaughtered (see Fig. 2.1). Feathers need to be removed from the body before the muscles around the feather papillae contract in rigor mortis. The poultry carcass can then be chilled and stored in this condition for several weeks. More commonly, the heads, necks and feet are removed from the carcasses and they are eviscerated (Fig. 2.2). The skin around the cloaca of the bird is cut and the inedible and edible viscera are removed through this vent. About one quarter of the body weight of poultry is inevitably lost during slaughter and evisceration (Table 2.1).

There are three main ways of marketing poultry meat; as an eviscerated whole carcass, as unboned commercial cuts or as boned poultry meat (Fig. 2.3). Most poultry meat is sold as an eviscerated carcass. This often includes the skin of the bird and some of its edible viscera.

Eviscerated carcass yields generally increase with increasing body weights. The digestive tract, respiratory tract and other inedible parts form a large part of the weight of a newly hatched bird. The weights of these parts decrease in proportion to the rest of the body as the birds grow. However, body fatness often increases when the birds near their mature body size. Some adipose tissue in the body cavity will be lost at

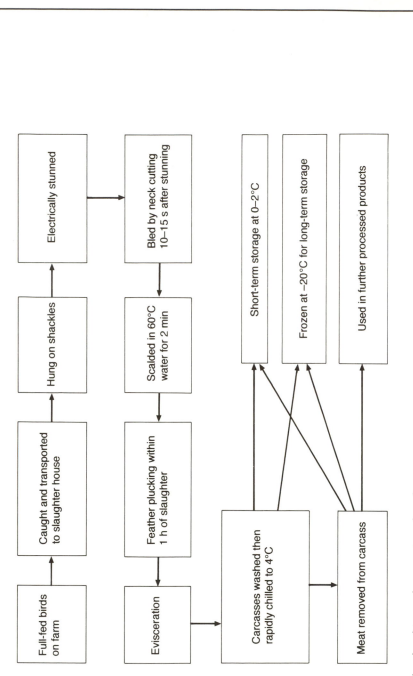

Fig. 2.1. The slaughter and processing of poultry.

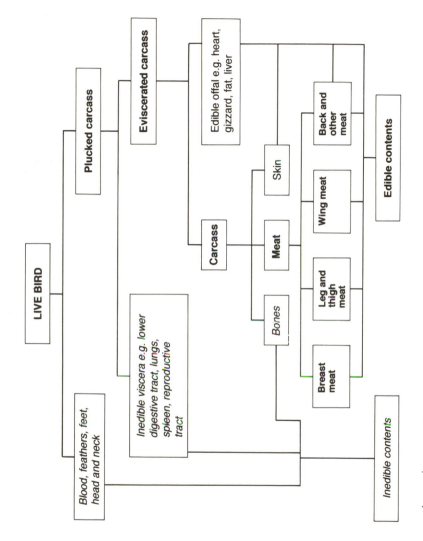

Fig. 2.2. The contents of a poultry carcass.

Table 2.1. Composition of poultry carcasses[1]

	Domestic fowl		Turkeys	Ducks		Geese
Carcass content	Meat-line strains	Egg-laying strains		Common ducks	Muscovy ducks	
(% of liveweight)	(1.8 kg broilers)	(1.7 kg laying hens)	(5.0 kg)	(2.7 kg)	(2.5 kg)	(5.5 kg)
Eviscerated carcass	73.7	66.5	78.9	71.2	71.4	72.9
Edible meat	42.7	37.1	51.3	28.8	36.3	34.3
Other edible parts	16.0	16.7	11.6	29.1	21.2	26.6

[1]Data adapted from Richter *et al.* (1989).

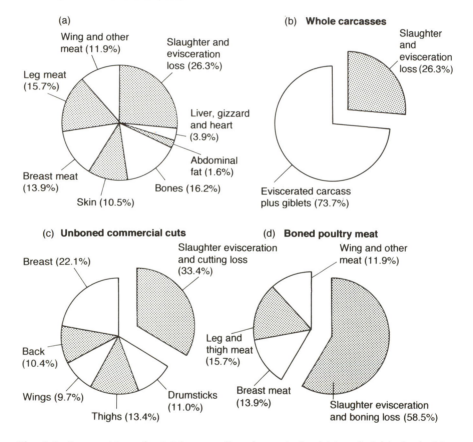

Fig. 2.3. Composition of a 1.8 kg meat-line domestic fowl (a) and yield of saleable parts depending on whether they are sold as whole carcasses (b), unboned commercial cuts (c) or boned poultry meat (d).

evisceration, so the percentage carcass yields decrease slightly at high carcass weights.

Table 2.2. Yields of poultry meat.[1]

Carcass content	Domestic fowl		Turkeys	Ducks		Geese
	Meat-line strains	Egg-laying strains		Common ducks	Muscovy ducks	
(% of total meat on carcass)	(1.8 kg broilers)	(1.7 kg laying hens)	(5.0 kg)	(2.5 kg)	(2.5 kg)	(5.5 kg)
Breast meat	33.5	31.5	38.0	31.4	33.7	37.1
Leg and thigh meat	37.9	37.6	31.2	29.8	30.2	28.8
Wing and other meat	28.8	30.8	39.0	38.1	38.1	34.3

[1]Data adapted from Richter *et al.* (1989).

Increasing amounts of poultry meat are now cut from the carcass before sale. The relative yields of these parts on a poultry carcass are therefore important (Table 2.2). Breast meat often has a higher economic value than meat from other parts of the poultry carcass. The percentage yield of breast meat increases with increasing body weight, so birds are often grown to heavy weights if meat is to be cut from the carcass (Fig. 2.4).

Meat

Muscles are needed by birds for all types of movement. There are three types of muscle: (i) red muscles, which are attached to the skeleton and comprise most of the muscles of the body; (ii) heart muscle, which has a similar structure to the red muscles; and (iii) white muscles, which occur in the walls of blood vessels and the digestive tract.

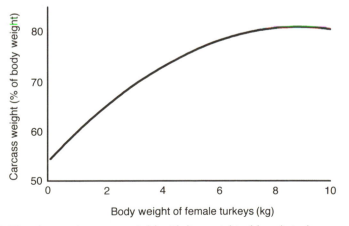

Fig. 2.4. The changes in carcass yield with liveweight of female turkeys.

The distribution of skeletal muscles in poultry is dominated by their original need for flight. The wings are forced downwards by pectoral muscles, commonly known as the breast muscle. The wings are then lifted by muscles that form most of the meat on the back. Strong leg muscles are needed for takeoff and landing, also walking, running and perching. Large extensor and flexor muscles are present in the upper part of the leg. A few muscles are present in the wings to control the orientation of the flight feathers.

Most species of poultry are unable to fly even though they may have been selected for large breast muscles. The breast muscles are only used for short bursts of flapping to escape a predator or another perceived danger. A poor blood supply and a low concentration of mitochondria are enough to meet these requirements, so these muscles fatigue very easily. The breast muscles also only need a low concentration of myo-globin, a dark red compound, hence breast meat has a lighter colour compared with leg meat.

Once a bird is slaughtered, its muscles start to contract. Oxygen soon runs out because there is no longer a blood supply. Only anaerobic metabolism can then take place and lactic acid builds up in the muscle. The concentration of ATP within the muscle steadily decreases until it disappears completely. Permanent actomyosin bonds rapidly form once ATP is depleted and the muscles stiffen. This is termed rigor mortis.

Quality (Table 2.3)
Blood removal

Most slaughter methods involve bleeding the birds. The exception is slaughter by neck dislocation. However, neck dislocation creates a void in the neck into which blood runs from the ruptured blood vessels.

Poultry lose about 45% of their total blood volume if they are properly bled. The rest of the blood stays with the carcass, particularly in the viscera. Improper bleeding may result in less than one third of the total blood volume being lost. The nutritional quality of a carcass is not affected by improper bleeding but the blood haemoglobin gives a redder colour to the meat and skin in these carcasses so the carcass has a poorer visual appearance. Engorged blood vessels, particularly near bones in the wings and legs, also detract from the appearance of the cooked carcass. Small haemorrhages in the muscle can cause 'blood speckling' of the cooked meat.

There are no major differences between bleeding methods in the total loss of blood from the carcass, but electrical stunning can increase the rate of blood loss. Alternating current (50 Hz giving at least 105 mA) for 7 seconds when the heads of the birds are immersed in a water tank is a suitable electrical current for bleeding. This also effectively stuns the birds. Neck cutting should be 10 to 15 seconds after stunning so that

Table 2.3. Chemical composition and product quality of poultry meat.[1]

	Domestic fowl		Turkey	
	Breast meat	Dark meat	Breast meat	Dark meat
Chemical composition (%)				
Water	74.4	74.5	75.2	75.9
Protein	21.8	19.1	23.2	20.3
Fat	3.2	5.5	1.1	3.6
Product characteristics				
Colour (reflectivity units)	27.6	16.5	43.5	27.8
(High values = light colour)				
pH (after 24 h)	5.8	6.4	–	–
Drip-loss (%) over 24 h	7.3	2.1	–	–
Cooking loss (%)	23.5	23.5	28.2	28.2
(20 min at 200°C)				

[1]Data adapted from Richter *et al.* (1989).

the heart rate can return to normal levels. This improves the efficiency of bleeding. A bleeding time of at least 90 seconds should be given for domestic fowls and this should be increased to at least 2 minutes for larger birds such as turkeys and geese. These delays also ensure that the birds are dead before they enter the scald tanks before plucking.

Acidity and rigor mortis

There is a drop in pH in the muscle as lactic acid builds up after slaughter. A low final pH improves the keeping quality of the meat by reducing bacterial growth. Chicken breast muscle usually drops to a final pH of 5.6–5.9 whereas leg meat drops to a pH of 6.1–6.4. Leg meat has a correspondingly poorer keeping quality compared to breast meat.

The final pH also affects the toughness of meat. A bird that has been under chronic stress before slaughter may have low reserves of muscle glycogen. The resulting meat may have a pH above 6.4 and will be dry, firm and dark coloured. The high pH does not allow enough protein degradation within the muscle and so the meat remains tough and unattractive.

The whole carcass must enter rigor mortis for the resulting muscle to have its usual tenderness. Poultry meat remains tough if it is cut from the carcass before rigor mortis has begun. The meat will also be tough if a whole carcass is frozen before rigor mortis is begun and then rapidly defrosted and cooked. This is called 'cold-shortening'.

The rate that a carcass reaches rigor mortis affects its quality. Pale, soft and exudative (PSE) meat is occasionally seen in chickens and turkeys, although it is more common in pigs. Poultry meat classified as PSE may reach an acceptable final pH, but the rate of drop of pH will have been excessively high. Excessively fast glycolysis results in the meat entering rigor mortis too quickly. There is not enough time for protein degradation so the meat remains tough. The decreased protein breakdown reduces the water holding capacity in the meat and so it exudes moisture.

Poultry enter rigor mortis quicker than other farm animals. A meat-line chicken will enter rigor mortis in around 1 hour and it will take 90 minutes in turkeys. A carcass that has PSE meat may have entered rigor mortis within 5 minutes of slaughter. A bird that struggles excessively before slaughter will have a high rate of post-mortem glycolysis. Mechanical plucking increases the rate of muscle glycolysis compared to manual plucking. The meat of mechanically plucked birds is slightly tougher than that of manually plucked birds.

The temperature of the muscle after slaughter of the bird can also affect the rate of entering rigor mortis. More than 20°C gives an excessive contraction of the muscle and a rapid depletion of glycogen. Many slaughter operations therefore cool the carcass to 4°C quickly after evisceration. The carcasses are cooled by air chillers or by immersion in refrigerated brine. The temperature of the muscle must not go below 0°C else 'cold-shortening' occurs. The muscle fibres lose calcium ions too quickly and this also gives an excessive contraction of the muscle. Cold shortened meat also tends to be tough.

Flavours and taints

The flavour of poultry meat is difficult to alter. Extensive or intensive production systems give poultry meat with a similar flavour. Similarly, the pre-slaughter handling of birds has little effect on the ultimate flavour of the meat.

Most dietary constituents neither improve nor impair the flavour of poultry meat. Fish oils are an exception. Birds fed diets with greater than 15 g kg^{-1} of fish oil have meat with a fishy taint because the long chain, unsaturated fatty acids present in fish oil transfer into the muscle.

Many attempts have been made to improve the flavour of poultry meat by putting additives in the birds' diets. Few attempts have been successful. For example, dietary garlic oil gives subtle changes in the flavour of the poultry meat. Taste panels are often undecided whether the changed flavour is an improvement or not.

Environmental contaminants are a frequent cause of tainted poultry meat. For example, chemical compounds may be present in wood preservatives used on the structure of poultry houses. These compounds may contaminate the air within the poultry house and so enter the birds' bodies by way of their respiratory tracts. Alternatively, these compounds may contaminate the wood shavings used as bedding material and the birds may then eat some wood particles. Some disinfectants that contain phenols may also cause taints. Innocuous chemical compounds can be converted to tainting compounds by the actions of fungal organisms, so the incidence of taints in meat can be sporadic and highly variable.

Poultry meat flavour can be improved if the poultry carcasses are hung before they are eviscerated. Birds can be slaughtered, chilled and held with their digestive tracts intact for up to 3 weeks without increasing the microbial contamination of the meat. The flavour of the meat becomes more intense with increased storage time. The flavour is often described as 'livery' and is similar to the flavour of game birds. The acceptability of the flavour depends on personal preferences. The development of the flavour is caused by leaching of flavour compounds from the digestive tract into the meat. The type of bacterial flora in the digestive tract can affect the development of these flavours.

Fat

Distribution on the carcass

Poultry, like mammals, store excess energy as fat. Much of this fat is stored in adipose tissues throughout the body. Adipose tissue is a matrix of cells, called adipocytes, that are filled with triglycerides. Adipose tissue contains about 80% fat and 20% water with a small amount of protein. Most adipose tissues are found either within the body cavity or under the skin of poultry.

The adipose tissue in the body cavity surrounds the upper and lower parts of the digestive tract and the kidneys. A further large depot is in the lower part of the body cavity near the cloaca. This is often called 'abdominal fat' or 'leaf fat'. Abdominal fat is about half the weight of the total fat in the body cavity and it can often account for 2% of the total body weight of a meat-line domestic fowl. Most of the abdominal fat will be left attached to the eviscerated carcass. This fat is usually considered part of the edible product.

There is also adipose tissue under the skin of birds. This subcutaneous fat can account for 2% of the total live weight of a meat-line domestic fowl. The main depots of this subcutaneous fat are on the neck and back, the breast, the legs and thighs. The fat on the neck and back is about three quarters of the total subcutaneous fat. Subcutaneous adipose tissue is strongly attached to the skin of birds and will remain with the skin if the carcass is skinned.

Fat can be found in all parts of the bodies of poultry. The fat contained in adipose tissue is only about one quarter of the total fat content of the carcass. Other fat-containing parts of poultry are as follows: poultry skin (40% fat), the skeleton (10% fat) and the liver (6% fat). The feathers contain only 2% fat in most species.

Quality
Fatty acid composition

Poultry fat has about 38% saturated fatty acid and 62% unsaturated fatty acids when the birds are fed fat-free diets. However, the diet of the birds changes the fatty acid composition of the adipose tissues. The saturated : unsaturated fatty acid ratio stays at 38 : 62 if saturated fats are fed. The ratio changes up to 22 : 78 if the diet contains mostly unsaturated fatty acids.

High unsaturated fat intakes may be preferable for humans. They raise blood cholesterol levels less than do saturated fats. However, unsaturated fatty acids are more prone to oxidation and high concentrations of oxidized fat cause carcass taints. Poultry diets that contain high levels of vitamin E, a natural antioxidant, can reduce the amount of fat oxidation in their stored carcasses.

Colour

Human populations around the world have different preferences for skin and fat colour in poultry carcasses. The colour may vary from a deep orange/yellow to white. This skin colour is related to the dietary intake of yellow, fat soluble compounds, called xanthophylls. Maize (15 mg kg^{-1}), maize gluten meal (200 mg kg^{-1}), grass meal (200 mg kg^{-1}) and lucerne (150 mg kg^{-1}) are rich natural sources of xanthophylls. Xanthophylls can be chemically synthesized or extracted from naturally rich sources, for example marigold petals. Xanthophylls are the same compounds that enhance the yellow colour of egg yolks.

A high xanthophyll intake gives a high xanthophyll content of all the fat in the body. The skin (40% fat) becomes yellow, as do other adipose tissues within the body. More than 10 mg of xanthophyll per kilogram of diet is usually satisfactory for meat birds for markets that prefer a yellow coloured carcass. A dietary xanthophyll content of less than 2 mg kg^{-1} is satisfactory for 'white' skinned carcasses.

Dietary xanthophylls have different potencies in colouring body fat in different poultry strains. Some breeds of domestic fowl are less efficient at converting dietary xanthophyll into xanthophyll in body fat.

Liver

In some developed countries there is a market for a high-value poultry product called foie-gras. Foie-gras is the liver of geese or ducks that contains up to 70% fat compared to conventional duck or goose liver that is usually less than 15% fat.

Foie-gras is produced by force-feeding geese or ducks large excesses of energy. A 2 or 3 week 'cramming' period is often used where a highly digestible carbohydrate source, usually steamed maize, is given. The feed is placed directly into the birds' crops up to six times a day. The carbohydrate is converted to fat in the liver but such a high rate of fat synthesis overloads the fat transport mechanisms from the liver. Fat accumulates and large fat globules fill the cytoplasm of the liver cells.

Cramming gives a disproportionate increase in the weight of the liver. Body weights can increase by 70% over a 3-week cramming period whereas liver weights can increase by over 600%. Livers of crammed ducks and geese may weigh over 500 g, about 10% of their body weight, when the birds are slaughtered.

EGGS

The egg is a complex structure distinguished by having four very different main parts; these are yolk, albumen, shell membranes and shell (Fig. 2.5).

The blastoderm is a small disc that contains the genetic code of the egg. The blastoderm lies on the surface of an almost spherical yolk. Yolk material is enclosed in a delicate, elastic membrane.

Two chalazae are firmly attached to the surface of the yolk membrane. They are positioned either side of the yolk. The chalazae are long twisted fibres. The other ends of the chalazae are interlaced with fibres in the albumen. Their purpose is to stabilize the position of the yolk and hold it near the geometric centre of the egg.

The albumen surrounds the yolk and acts as a shock absorber also a supply of some nutrients. There are three layers of albumen (Fig. 2.6). A liquid albumen surrounds the yolk. This accounts for one-sixth of the total albumen. A middle layer contains a thicker albumen that has a

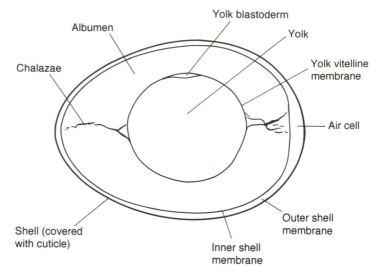

Fig. 2.5. The structure of the egg.

higher concentration of mucin fibres. This layer accounts for 57% of the total albumen in the eggs of domestic fowl. The outer albumen layer is next to the shell and is similar in composition to the inner albumen layer. The two different albumen types are often described as 'thick' and 'thin' albumen.

Two membranes separate the albumen from the shell. Each membrane consists of a matrix of fibres that allow gases and liquids to diffuse through. The volume of water lost from the egg during storage is replaced by air. This air is stored in the air cell sandwiched between the inner and outer shell membranes at the blunt end of the egg. The size of the air cell is often taken as a measure of the length of storage of the eggs. Many factors can affect the loss of water vapour from an egg, so this 'freshness indicator' is not reliable.

Egg shells perform a difficult task. They have to be strong and rigid to protect the developing embryo and yet porous to allow gaseous exchange with the outside air. The shell strength has to be enough to

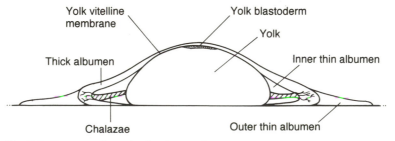

Fig. 2.6. Cross-section of a 'broken-out' egg.

support the weight of an adult yet weak enough to allow a chick to hatch. The shell thickness is generally related to the size of the egg and the size of the adult bird.

Yields of Egg Components

A domestic fowl's egg contains approximately 64% albumen, 27% yolk and 9% shell. The chalazae (0.25% of the total egg weight) are usually included with the albumen weight. Shell membranes (0.75% of the total egg weight) are generally included with the shell weight. The species differences in the weight of the three main components of eggs are shown in Table 2.4.

There are two major factors that can affect the yields of components from eggs.

1. Heavier eggs have proportionately less yolk and more albumen. The proportion of shell in the egg does not change with egg weight.
2. The duration of the laying period also changes the yield of egg components. As a bird goes further into its laying period its eggs have more yolk and less albumen. These changes are still evident if the weights are adjusted to a constant egg weight. The proportion of shell in the egg also declines as birds get older. Some changes are reversed if the birds moult and enter a second laying period.

Birds lay heavier eggs as they get older. The changes in yolk and albumen composition due to increased egg weight with age are partly counteracted by the changes due to the increasing length of the laying period. Egg composition does not, therefore, change greatly through the laying period of an individual flock.

Table 2.4. The composition of eggs from different poultry species.[1]

Species	Egg weight (g)	Egg components (% of total egg weight)		
		Albumen	Yolk	Shell
Domestic fowl[2] (egg-laying strains)	57	63.8	27.2	9.0
Domestic fowl[3] (meat-bird strains)	61	59.4	29.1	11.5
Turkey[4]	89	57.4	32.6	10.0
Guinea fowl	40	52.3	35.1	12.6
Quail[5]	13	59.5	32.2	8.3
Common duck	65	52.6	35.4	12.0
Goose	130	52.5	35.1	12.4

[1]Data from Romanoff and Romanoff (1949) except for [2]Curtis *et al.* (1986), [3]O'Sullivan *et al.* (1991), [4]Nestor *et al.* (1982), and [5]Rose (1995) Personal data.

Table 2.5. The nutrient composition of eggs from different poultry species.[1]

Species	Egg weight (g)	Water (%)	Protein (%)	Lipid (%)	Carbohydrate (%)	Minerals (%)	Cholesterol (mg 100 g[-1])
Domestic fowl	57	74.6	12.1	11.1	1.2	1.0	548
Turkey	79	72.5	13.7	11.9	1.1	0.8	933
Quail	9	74.3	13.1	11.1	1.4	1.1	844
Common duck	70	70.8	12.8	13.8	1.4	1.2	884
Goose	144	70.4	13.9	13.3	1.3	1.1	–

[1]Data adapted from Posati and Orr (1976).

Composition and Quality of Eggs

Whole eggs
Content of major nutrients

Eggs are a concentrated source of nutrients compared to the amount of energy they supply. The yolk is particularly rich in nutrients. Yolk supplies approximately 50% of the protein in the egg and nearly all the fat. The nutrient composition of eggs from different poultry species is shown in Table 2.5.

Egg protein is highly digestible. The levels of amino acids are similar to the balance of amino acids needed by men and women. The fat within the egg is emulsified and is highly digestible. It contains high levels of unsaturated fatty acids although this is affected by the diet of the birds.

The mineral compositions of the edible parts of eggs are relatively high. Eggs are particularly rich in iron and phosphorus. Mineral compositions can also be altered by the diets given to the laying birds. Eggs contain all vitamins except vitamin C. They are particularly rich in the fat-soluble vitamins A and D.

Cholesterol

Eggs have a high concentration of cholesterol. Each domestic fowl egg contains around 270 mg of cholesterol and a duck egg contains 620 mg. An adult man living in a developed country and eating a mixed diet would eat about 450 mg of cholesterol daily. A high egg intake can thus substantially increase the cholesterol intake of a person.

There is a correlation between blood cholesterol concentrations and the risk of coronary heart disease. Coronary heart disease accounts for a quarter of all deaths in most industrial countries. Populations with a 1% lower blood cholesterol have a 1–2% lower incidence of heart disease. A blood cholesterol concentration that is more than 5.2 mmol l[-1] (20 mg dl[-1]) may give an individual a higher risk of coronary heart disease although there is much conflicting evidence. The incidence of heart disease is a combination of dietary, environmental and other risk factors. There are at least seven dietary factors that can affect blood cholesterol levels or the incidence of coronary heart disease. A high cholesterol intake is just one of these factors.

Most of the cholesterol within the body is manufactured in the liver. A high dietary cholesterol intake results in a reduction in cholesterol synthesis within the body. A high dietary cholesterol intake does not generally increase blood cholesterol as much as expected. However some people do not reduce liver cholesterol synthesis when they have high cholesterol intakes. This small group of people is most at risk from maintaining a high dietary cholesterol intake.

There have been many studies aimed at reducing the cholesterol contents of eggs. Genetic selection has been unsuccessful in developing a low cholesterol strain of egg laying poultry. Diets that are very low in cholesterol or high in plant sterols may reduce egg cholesterol by up to 25%. Diets that are high in fibre can reduce egg cholesterol by up to 10%. Some chemicals reduce egg cholesterol but none allow good egg outputs with residue free eggs.

Microbial contamination

Eggs are a rich source of nutrients and an ideal medium for bacterial growth. Broken-out eggs can quickly become contaminated with pathogenic organisms. Many other common foods have a similar risk of pathogenic bacterial contamination.

Intact shell eggs are relatively free of bacteria. Egg shells and their membranes provide a barrier against bacterial invasion. The albumen contains antimicrobial substances that attack any bacteria that do enter. The main risks of bacterial contamination occur if the eggs are broken out, or if the shell is cracked, or if there is excreta contamination on the shell.

Food poisoning outbreaks caused by *Salmonella enteritidis* began to increase sharply in the 1980s. Some of these outbreaks were traced to the consumption of eggs, or egg-containing products. The eggs had been intact and of high quality. It was then realized that a new strain of *S. enteritidis* had developed. This was called phage type 4 (PT4). This *Salmonella* strain can infect the ovaries of laying hens and cause them to lay about 0.5% of their eggs with a few *Salmonella* organisms contained inside.

Salmonella enteritidis PT4 organisms are usually found on the surface of the yolk membrane in a newly laid egg. If the organisms remain at this site they do not multiply because the albumen has an antimicrobial affect that limits growth. The bacteria will also not multiply if they migrate in to the main body of the albumen. If the microbes enter the yolk they can quickly multiply if the egg temperature is greater than 8°C.

A whole egg in which there has been a rapid multiplication of *S. enteritidis* organisms can be a cause of a food poisoning outbreak. This is particularly a problem if the egg is eaten raw but mild cooking may not kill *S. enteritidis* within the yolk. The high fat content and the viscosity of the yolk protects the *Salmonella* cells from the effect of heat. Cooked egg products, for example scrambled eggs, could still be contaminated with *S. enteritidis* organisms. The risk of food poisoning from infected eggs can be very low. Only 1 in 14,000 eggs contain *S. enteritidis* PT4 organisms in areas where the infection of laying hens is endemic.

The spread of *S. enteritidis* PT4 infections is difficult to control because, although the disease is mainly spread by infecting the ovaries of embryos in infected eggs, there are also other routes of infection: *S. enteritidis* PT4 infection can invade the blood system and then infect the ovaries. Organisms in the digestive tract can be transported in the blood to the ovaries and other body organs. Contamination of the bird's feed or environment can result in infection of laying birds. *Salmonella enteritidis* is thus difficult to eliminate from egg laying flocks and programmes that test for *S. enteritidis* and then slaughter infected flocks have not successfully eradicated the disease from countries or areas.

High temperature storage of eggs allows rapid multiplication of *Salmonella* organisms. Food poisoning outbreaks peak in warmer summer months. Refrigeration of shell eggs reduces the risk of food poisoning outbreaks. Poor food hygiene practices further increase the risk of a food poisoning outbreak (Fig. 2.7).

Yolk

Yolk consists of alternate layers of dark and light coloured yolk material. A transparent vitelline membrane surrounds and contains the yolk. Dark yolk contains 45% water whereas light yolk consists of 86% water. The layers of light and dark coloured yolk material can be seen in a complete yolk but they are almost impossible to separate.

Yolk consists of a suspension of particles in a protein solution. Livetin, a protein that is rich in sulphur amino acids, is the main component of the protein solution. Vitellin is a phosphorus-containing protein that is the major protein of the suspended particles. Much of the vitellin is complexed with lipids to form lipovitellins. Phosvitin is a non-lipid phosphoprotein that is also present.

The yolks of eggs from all poultry, except waterfowl, contain about 49% water, 16% protein and 33% fat. Two-thirds of the fat in the yolk is present as triglycerides. There are 30% phospholipids and 5% cholesterol. Yolks of waterfowl eggs have higher fat (36%) and protein (18%) content with correspondingly less water (44%).

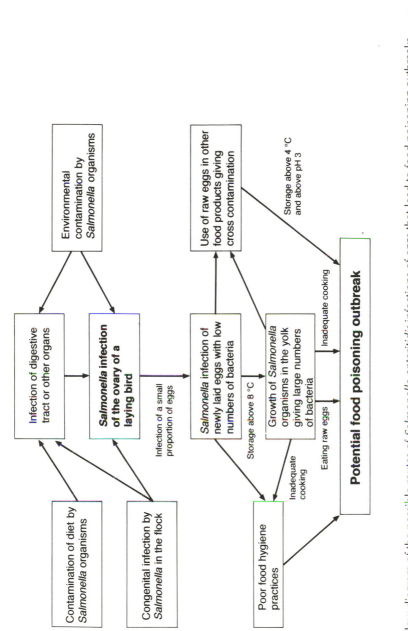

Fig. 2.7. Flow diagram of the possible routes of *Salmonella enteritidis* infections of eggs that lead to food poisoning outbreaks.

Yolk gains water from the albumen during the storage of whole eggs. The water content of the yolk may vary from 46% to 50% depending on the length and conditions of storage.

There are few influences on the protein composition or amino acid composition of the yolk. Similarly, the total lipid content of the yolk material is relatively constant.

Diet can change the fatty acid composition of the lipid. The proportion of saturated fatty acids, mainly palmitic and stearic acid, does not change with diet. Their levels remain between 30–38% of the total lipid. A diet that is high in polyunsaturated fatty acids (PUFAs) increases the PUFA content of the egg lipid. Oleic acid, a monounsaturated fatty acid, is usually reduced in high PUFA yolk lipid.

Quality

Blood spots and meat spots   Some eggs are laid with a spot of blood underneath the vitelline membrane. A variable amount of blood may be present. Most marketing regulations do not allow these eggs to be sold as first quality eggs. They are nutritionally safe for consumption, but there is often much consumer dislike of this quality defect. Blood spots are caused by the rupture of a small blood vessel while the ovum is still attached to the ovary. Different strains of birds can differ in their incidence of blood spots and some management factors can increase their incidence.

Meat spots are brown or white spots that are found on the surface of the vitelline membrane. Meat spots may be caused either by a fragment of shell, including the brown porphyrin pigment, or by the presence of some degenerated blood. There are no factors known to influence the incidence of meat spots.

Colour   The intensity of the yellow colour of yolk can be easily altered. Different markets around the world have different preferences for the intensity of yellow. The yellow pigments in yolk are the same compounds that give the yellow coloration to poultry fat.

The yellow colour of the yolk is caused by lipid-like compounds, called xanthophylls. The xanthophyll content of the yolk is almost completely dependent on the xanthophyll content of the bird's diet. Grass has a high xanthophyll content (20 mg kg^{-1}) so birds kept on range may eat enough grass to give their eggs a dark yellow colour. Maize is particularly high in xanthophylls (15 mg kg^{-1}), so feeds based on maize typically give a deep yellow coloration to egg yolks. Most other cereals, such as wheat, barley or rice, contain no xanthophylls. Diets based on these cereals need to be supplemented with a concentrated source of xanthophylls if a deep yellow yolk is wanted.

Different xanthophylls give different intensities of yellow. For example, lutein gives a lemon-yellow colour whereas zeaxanthin gives a golden-yellow colour. Rich yellow colours can only be achieved by using combinations of xanthophylls.

Albumen
Composition

Albumen is the main water store of the egg and it contains about 88% water. The albumen is a protein solution in which there are long fibres of a protein, ovomucin. Ovomucin accounts for about 75% of the total protein content of the albumen. Two globular proteins account for a further 20% of the total albumen proteins. The albumen contains an enzyme, lysozyme, which attacks the cell walls of any invading bacteria. There is only about 1% total carbohydrate in the albumen and it contains almost no lipid. All species of poultry have a similar albumen composition. Layers of thick and thin albumen surround the yolk of the egg. The water content is not directly related to the thickness of the albumen. Thick albumen has four times the concentration of ovomucin than the thin albumen which is probably the main cause of its thicker gel-like structure.

Quality changes during storage

Egg albumen loses water as the egg is stored. Most water is lost through the shell although some water goes into the yolk. Air moves into the air cell to replace the loss of volume of water. The size of the air cell is a good indicator of the amount of water loss. There is also a loss of carbon dioxide from the egg on storage. This gives an increase in the pH of the albumen. Albumen from a freshly laid egg will have a pH of 7.6 and this can rise to 9.2 after 14 days of storage. Both water loss and pH rise are dependent on the temperature and humidity during storage of the eggs.

The viscosity of the albumen decreases as eggs are stored. An interaction between ovomucin and lysozyme during storage is probably the major reason. Thin egg whites are sometimes a problem in marketing eggs. Consumers may associate thin albumen with excessively long egg storage. The excessive spreading of the egg white when the egg is broken out may be disliked for some methods of food preparation. This characteristic is assessed by measuring the height of the thick albumen layer. Albumen height is then corrected for egg weight and expressed as Haugh units. High storage temperatures increase the rate of decrease in Haugh units.

Shell (Fig. 2.8)

Composition

Egg shells are composed of 98% calcite (calcium carbonate) crystals and 2% protein. There is also a small amount of magnesium and phosphorus in the shell. The shell thickness varies from 0.13 mm in quail eggs to 0.45 mm in turkey eggs.

Protein in the shell combines with polysaccharides to form a matrix. Small parts of this matrix are bedded in the outer shell membrane and calcite crystals are embedded within it. The crystals are arranged randomly but they make a cone shape. In domestic fowl eggs the cone layer is about one third of the thickness of the whole shell (4 mm) and each cone is about 0.1 mm in diameter.

Superimposed on top of each cone is a tall column of long calcite crystals. The crystals are formed around a matrix of fine protein fibres. The columns are closely attached to each other. They form a layer that forms the greatest thickness of the egg shell. This area of elongated columns is called the palisade layer.

The palisade layer is closely packed together. Occasionally an air space occurs between some columns. This forms a channel through the egg shell and forms an oval pore on the surface of the egg shell. It forms a route for gaseous exchange though the egg shell. These pores in domestic fowl eggs are variable in size but on average they are about 20 μm in diameter. There are about 7500 pores on an egg from a domestic fowl.

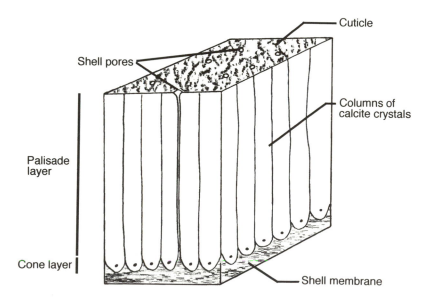

Fig. 2.8. A diagrammatic cross-section of an egg shell.

A fluffy, water-insoluble cuticle covers the surface of the egg. The cuticle protects the shell from damage, but it also partly blocks the pores and impedes entry by bacteria.

Quality

Strength   About 7% of all eggs have some degree of shell damage before they reach the consumer. Whole eggs can be subjected to large amounts of collision damage during laying, collection and transport. Some of these collisions will cause the egg to break. However, certain characteristics of egg shells make some eggs more liable to break.

The thicker the egg shell the stronger it is, and eggs that have a large diameter relative to their length are stronger. Shell thickness and egg shape only explain about 75% of the variation that exists in shell breaking strength. Shell strength is reduced when the formation of the palisade layer and, especially, the cone layer is disrupted. Cones that are compact with a small diameter relative to the height of the palisade layer cause the shell to have a high resistance to breakage.

There is a large variation in shell strength between strains and between individuals within a strain. Guinea fowl have particularly strong egg shells. These eggs have a smaller cone diameter and therefore a higher column density in the palisade layer compared to the eggs of the domestic fowl.

Shell strength can be affected by management factors. The average shell strength of eggs of an individual laying bird decreases as its laying period increases. Diets that are deficient in calcium, or vitamin D, result in thin egg shells. Lighting programmes, such as long ahemeral cycles, give thicker egg shells. These changes in shell strength and the estimation of shell strength in flocks are explained in more detail in Chapter 4.

Colour   Egg shell colour can vary considerably within poultry species. Domestic fowl strains lay eggs that vary in shell colour from white to a rich dark brown. Turkey eggs vary in the degree of brown speckling. Quail eggs are characteristically speckled with dark brown patches although some strains lay white eggs. Some duck eggs may have a green tinge on the shell.

The shell colours are mostly produced by pigments called porphyrins. Porphyrins are cyclic compounds secreted from the oviduct. Biliverdin, a bile breakdown product, is present in shells that have a green hue.

Shell pigments only occur in the outer part of the egg shell. They are irregularly embedded in the calcite crystals in the outer part of the palisade layer. The pigments are secreted from the oviduct only in the later stages of shell deposition. The pigment secretion may be localized so that the egg shell gets a speckled colour.

All strains of domestic fowl secrete some porphyrins into the egg shell and porphyrin is present in small amounts even in white shelled eggs. Shell colour differences between poultry strains are due to the different concentrations of porphyrins. There are distinct preferences between consumer groups around the world for particular shell colours. There are no major differences in egg composition or quality between white and brown eggs. Shell defects and internal egg defects such as bacterial contamination, meat spots and blood spots are usually detected by holding the egg in front of a bright light. This is called candling. It is easier to detect these egg quality problems when candling white shelled eggs than with brown eggs.

FURTHER READING

Poultry Meat

Boorman, K.N. and Wilson, B.J. (eds) (1977) *Growth and Poultry Meat Production*. Proceedings of the 12th Poultry Science Symposium. British Poultry Science Ltd.

Demby, J.H. and Cunningham, F.E. (1980) Factors affecting the composition of poultry meat. A literature review. *World's Poultry Science Journal* 36, 25–66.

Jensen, J.F. (1982) Quality of poultry meat – an issue of growing importance. *World's Poultry Science Journal* 38, 105–113.

Leclercq, B. and Whitehead, C.C. (eds) (1988) *Leanness in domestic birds*. Butterworths, London.

Mead, G.C. (ed.) (1989) *Processing of poultry*. Elsevier Applied Science, London.

Mead, G.C. and Freeman, B.M. (eds) (1980) *Meat Quality in Poultry and Game Birds*. Proceedings of the 15th Poultry Science Symposium. British Poultry Science Ltd.

Mellow, C. and Diprose, R.J. (1987) Poultry composition and human health. *Proceedings of the Nutrition Society of New Zealand* 12, 60–66.

Mulder, R.W.A.W., Scheele, C.W. and Verkamp, C.H. (eds) (1981) *Quality of Poultry Meat*. Proceedings of the 5th European Symposium. Spelderholt Institute for Poultry Research, Beekbergen.

Stadelman, W.J., Olson, V.M., Shemwell, G.A. and Pasch, S. (1988) *Egg and Poultry-Meat Processing*. Ellis-Horwood Ltd, Chichester, UK.

Eggs

Hargis, P.S. (1988) Modifying egg yolk cholesterol in the domestic fowl – a review. *World's Poultry Science Journal* 44, 17–29.

Humphrey, T.J. (1990) Public health implications of the infection of egg-laying hens with *Salmonella enteritidis* phage type 4. *World's Poultry Science Journal* 46, 5–13.

Karunajeewa, H., Hughes, R.J., McDonald, M.W. and Shenstone, F.S. (1984) A review of factors influencing pigmentation of egg yolks. *World's Poultry Science Journal* 40, 52–65.

Morris, G.K. (1990) *Salmonella enteritidis* and eggs: assessment of risk. *Dairy, Food and Environmental Sanitation* 10, 279–281.

Powrie, W.D. and Nakai, S. (1986) The chemistry of eggs and egg products. In: Stadelman, W.J. and Cotterill, O.J. (eds) *Egg Science and Technology*. 3rd edn, Macmillan Publishers, Basingstoke, England.

Sim, J.S. and Nakai, S. (eds) (1994) *Egg Uses and Processing Technologies*. CAB International, Wallingford.

Solomon, S.E. (1991) *Egg and Eggshell Quality*. Wolfe Publishing Limited, London.

Ulbricht, T.L.V. and Southgate, D.A.T. (1991) Coronary heart disease: seven dietary factors. *The Lancet* 338, 985–992.

Wells, R.G. and Belyavin, C.G. (eds) (1987) *Egg Quality – Current Problems and Recent Advances*. Poultry Science Symposium No. 20. Butterworths, London.

Growth **3**

THE PHYSIOLOGY OF GROWTH

A healthy, well-fed bird increases its weight from the time it hatches until it reaches its mature body size. Growth involves an increase in the number of cells in the body and an increase in the size of the individual cells.

There are four main components of growth.

1. There is an increase in muscle weight. Muscle is composed mainly of protein and water.
2. There is an increase in the size of the skeleton that must support the muscle growth. Minerals, particularly calcium, are the main components of bones.
3. There is an increase in total body fat in adipose tissue. The adipose tissue is composed primarily of triglycerides with a small amount of water.
4. Feathers, skin and internal organs also increase in size as birds grow. This only accounts for a small proportion of the weight gain in meat-line birds.

These components have different mechanisms of growth. The following sections give brief descriptions of how this growth occurs.

Muscle Growth

Muscle growth is an increase in the thickness and the length of the existing muscle fibres. Muscle fibre numbers are mostly determined before a bird hatches. The numbers of fibres are unlikely to increase from shortly after hatching until the bird reaches its mature body weight.

The existing muscle fibres become thicker during growth by their myofibrils dividing and multiplying. Additional sarcomeres are added to the end of the myofibrils to increase the lengths of the muscle fibres. The numbers of nuclei in the muscle fibres increase, but the increase is not proportional to the muscle size as it grows. Muscle fibres are thicker in males than in females. Strains of birds within a species that have high meat yields have thicker muscle fibres than lower meat yield strains.

Muscle growth relies on there being protein deposition. A small amount of muscle protein is being continually broken down and lost and so there must be a net gain of protein in the muscle. There must be an adequate dietary protein intake for any muscle growth to occur and a protein deficiency reduces the rate of muscle growth of a bird. There is also a complex interaction of hormones within the body that affect the rate of muscle growth.

Bone Growth

Bone has two functions. It forms a rigid skeleton (Fig. 3.1) to support the musculature of the body and it acts as a reserve of calcium and phosphate in the body.

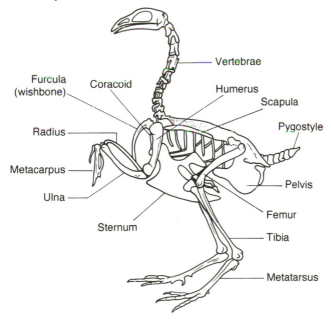

Fig. 3.1. The skeleton of the domestic fowl.

Bone is composed of a matrix of two phases. There is a mineral phase of calcium and phosphate. This phase is rigid and resists compression. The second phase is a network of organic fibres that are more able to withstand tension and torsion.

Bones grow in length at flat pieces of cartilage, called growth plates. At the growth plates are cells, called osteoblasts, that synthesize and secrete osteoid, a collagen-rich protein. Osteoid forms a matrix on to which calcium and phosphate ions are adsorbed and then crystals are formed. Osteoblasts are present elsewhere in the bone so that the cross-sectional area of the bones can also increase.

Osteoclasts are large cells that can be found throughout the bone, and they resorb both the mineral and organic phases in the bone. Bone resorption is needed to increase the size of the internal bone spaces. Resorption also provides a store of metabolically active calcium in the body. This is particularly important in laying birds because they need larger calcium reserves for egg shell formation. Bone growth is mostly controlled by genetic factors, but circulating hormones are also involved. Vitamin D and vitamin A deficiencies can also seriously reduce bone growth.

Leg injuries or deformities are a major problem in modern strains of fast-growing meat-line poultry. Skeletal deformities, caused by abnormal bone growth, are an underlying cause of many of these problems. The incidence of skeletal deformities is highly variable and it depends on the poultry strain and their age. Up to 8% of the birds in some flocks of meat-line domestic fowl suffer from leg skeletal deformities. The causes of these deformities are often complex and only a few bird management or dietary factors are known that can affect the incidence of these bone abnormalities (see Box 3.1).

Fat Growth

Fat growth occurs at many different sites in the body. It is primarily a store of energy but it has several other functions. For example, it acts as a form of body insulation. Most body fat is deposited in adipose tissue. The adipose tissue develops as pads at many distinct sites around the body (see Chapter 2).

Adipose tissue consists of several cell types, but adipocytes are the most numerous. Adipocytes are round or oval cells that accumulate triglycerides. Each adipocyte has a large droplet of lipid at the centre of the cell. The adipocytes are grouped together by a loose web of collagen fibres.

Box 3.1. Major non-infectious disorders of bone growth.

Spondylolisthesis

The spinal cord becomes trapped due to a deformation of one of the thoracic vertebrae. It may cause a twisted back and a rear paralysis that results in 'hock-sitting' or general immobility. It occurs mostly in meat-line domestic fowl.

Torsional or angular leg deformation

The intertarsal joint or the tibiotarsal joint of the leg becomes deformed. This gives an inward or outward turning of the leg. The shaft of the tibiotarsus bone may twist. The leg often rotates 90° to 180° in severe cases and it results in ligament rupture. Usually just one leg, most often the right leg, is affected. It occurs in turkeys, domestic fowl and guinea fowl.

Dyschondroplasia

There is an excessive swelling of the ends of the tibial bone of the leg. An abnormal mass of cartilage develops below the growth plate. It may give an abnormal gait or bowing of the legs and ultimately affects the bird's mobility. It occurs in domestic fowl, turkeys and ducks.

Rickets

Bone mineralization is inadequate due to a dietary deficiency of vitamin D, calcium or phosphorus. All bones soften and break easily. The leg bones bow with the body weight. It can occur in all species of growing poultry.

Each fat pad has its own characteristic number of adipocytes. This results in fat being accumulated most at some parts of the body. However, fatness can also vary due to differences in the size of the individual adipocytes. New adipocytes are formed early in life but the size of the lipid droplet inside the existing adipocytes can increase. This type of fat growth occurs more commonly as the bird nears its mature body size.

Fat is stored when a bird has a greater energy intake than its requirements. Fatty acids are accumulated in the adipose tissue from the blood stream. The fatty acids are derived directly from the diet or they are synthesized in the liver using glucose as a precursor. Poultry, unlike mammals, cannot synthesize fatty acids in their adipose tissue.

Fat is mobilized from the adipocytes in periods of energy deficiency. The hormone, glucagon, has the main influence on the rate of breakdown of the stored triglyceride.

MODELS OF GROWTH

Whole Body Growth

The weight of a growing bird increases towards a stable mature weight. All poultry species, and strains within a species, have characteristic mature body weights. Males and females of the same strain also have different mature body weights. This is called sexual dimorphism and the differences can be large in some species such as turkeys and Muscovy ducks.

Two poultry strains could have the same mature body weight but their growth may differ because they reach maturity at different ages. The rate of growth, or time taken to reach mature body weight, is another variable that describes the growth of a bird.

The rate of growth of a bird changes as it moves towards its mature weight. These changes give the bird a growth curve that has a characteristic S shape (Fig. 3.2). The maximum rate of growth (g of body weight gain per day) often occurs when the bird has reached a quarter to a half of its mature weight. The age that maximum growth rate is reached is another variable that describes the shape of the growth curve.

Several equations have been used to describe the S-shaped growth curve. The Gompertz equation (Box 3.2 and Fig. 3.3) is now most frequently used although others may be just as valid.

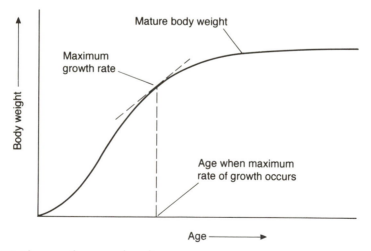

Fig. 3.2. The growth curve of poultry.

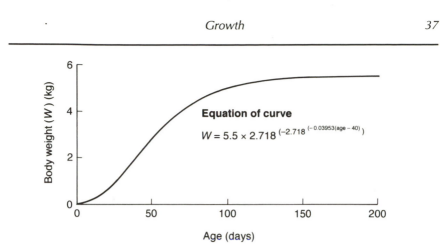

Fig. 3.3. The predicted growth curve of a male meat-line domestic fowl.

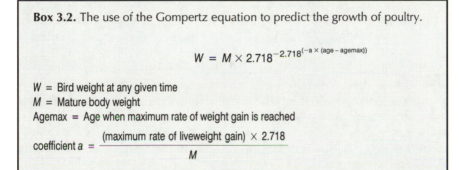

Box 3.2. The use of the Gompertz equation to predict the growth of poultry.

$$W = M \times 2.718^{-2.718^{(-a \times (age - agemax))}}$$

W = Bird weight at any given time
M = Mature body weight
Agemax = Age when maximum rate of weight gain is reached

$$\text{coefficient } a = \frac{(\text{maximum rate of liveweight gain}) \times 2.718}{M}$$

Table of coefficients to describe the growth curves of different poultry species.

Species	M (mature body weight) (kg)	Maximum rate of liveweight gain (kg day^{-1})	Coefficient a	Agemax (days)
Male meat-line domestic fowl	5.5	0.080	0.03953	40
Female meat-line domestic fowl	3.8	0.060	0.04290	35
Female egg-line domestic fowl	1.9	0.017	0.02433	55
Male heavy strain turkey	16.8	0.0130	0.02103	85
Female heavy strain turkey	10.8	0.099	0.02492	69
Male Muscovy duck	5.0	0.080	0.04349	37
Female Muscovy duck	3.0	0.052	0.04711	32

Growth of Body Parts

The gain in weight of a bird is composed of increases in weight of its body parts. These body parts grow at different rates. For example, the weight of the liver of a meat-line domestic fowl is 4% of its body weight at 7 days but less than 2% at 7 weeks old. The weights of the digestive tracts and other visceral organs typically decrease in proportion to body weights as birds grow. Conversely, the weights of muscles and body fat typically increase as a proportion of body weight during growth.

Each body part has its own characteristic growth curve that can be described by a Gompertz curve. The sums of all these Gompertz curves add up to approximately give the growth curve of the whole animal. However, large series of growth curves are complicated to deal with.

Allometric growth ratios are used to describe the changes in growth of an individual body part compared to overall growth. These relationships are usually not linear but become linear if a logarithmic transformation is made on the data. The equation of this line is as follows:

$$\log (y) = \log (a) + k \log (x) \tag{3.1}$$

where: x = body weight (kg); y = weight of body part (kg); k = allometric growth ratio (slope of line); and a = constant (intercept). The linear regression of log body weight (x axis) against log weight of the body part (y axis) usually gives a highly significant correlation coefficient. The slope of the line, k, shows the relative growth of the body part

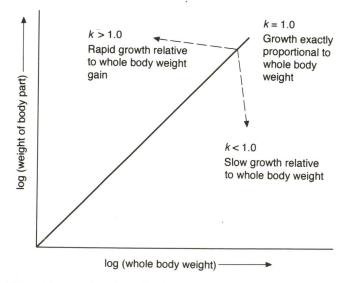

Fig. 3.4. Allometric growth ratios of body parts.

Table 3.1. Estimates of the allometric growth rates (*k*) of body parts compared to the total body weight gains of male birds of three poultry species (meat-line strains).

Body part	Domestic fowl	Common duck	Turkey
Liver	0.82	0.87	0.84
Digestive tract	0.59	0.90	0.76
Eviscerated carcass	1.10	1.11	1.09
Body fat	1.25	1.34	1.21
Body protein	1.03	0.99	1.05
Breast meat	1.21	1.37	1.18
Wings	1.14	1.01	1.06
Legs (drumsticks and thighs)	1.14	1.22	1.04

(Fig. 3.4). Some allometric growth ratios are shown in Table 3.1. Different strains within a species may have different ratios. Male and female birds within a strain have similar allometric growth ratios except for the growth of fat. Females deposit greater amounts of body fat at heavy body weights than males. The rate of deposition of body fat is also highly dependent on the diet given to the birds.

FURTHER READING

The Physiology of Growth

Loveridge, N., Farquharson, C. and Scheven, B.A.A. (1990) Endogenous mediators of growth. *Proceedings of the Nutrition Society* 49, 443–450.

Scanes, C.G. (1987) The physiology of growth, growth hormone, and other growth factors in poultry. *Critical Reviews in Poultry Biology* 1, 51–105.

Swatland, H.J. (1989) Physiology of muscle growth. In: Nixey, C. and Grey, T.C. (eds) *Recent Advances in Turkey Science*. Poultry Science Symposium No. 21. Butterworths, London, pp. 167–182.

Swatland, H.J. (1984) *Structure and Development of Meat Animals*. Prentice-Hall Inc., New Jersey.

Whitehead, C.C. (1992) *Bone Biology and Skeletal Disorders in Poultry*. Poultry Science Symposium No. 23. Carfax Publishing Company, Abingdon, UK.

Vernon, R.G. (1992) Control of lipogenesis and lipolysis. In: Boorman, K.N., Buttery, P.J. and Lindsay, D.B. (eds) *The Control of Fat and Lean Deposition*. Butterworth-Heinemann, London, pp. 59–81.

Models of Growth

Emmans, G.C. (1988) Genetic components of potential and actual growth. In: *Animal Breeding Opportunities*. British Society of Animal Production Occasional Publication No.12. BSAP, Edinburgh, pp. 153–181.

Emmans, G.C. (1989) The growth of turkeys. In: Nixey, C. and Grey, T.C. (eds) *Recent Advances in Turkey Science*. Poultry Science Symposium No. 21. Butterworths, London, pp. 135–166.

Kwakkel, R.P., Ducro, B.J. and Koops, W.J. (1993) Multiphasic analysis of growth of the body and its chemical components in White Leghorn pullets. *Poultry Science* 72, 1421–1432.

Parks, J.R. (1982) *A Theory of Feeding and Growth of Animals*. Springer-Verlag, Berlin.

Female Reproduction 4

THE REPRODUCTIVE TRACT

Poultry species can produce large numbers of eggs during their lifetime. Each egg contains one ovum and may be fertilized by male sperm inside the body of the female. The albumen and shell are then deposited round the yolk of the fertilized egg after fertilization.

The female bird continues to lay the same number of eggs whether or not she has been inseminated with sperm. Infertile eggs have the same chemical composition as fertile eggs. Edible egg production systems have lower costs if there are no males in the flock, so most eggs produced for human consumption are infertile.

Most laying birds cannot produce more than one egg each day. A sequence of days on which an egg is laid is broken by one day with no egg. Another daily sequence of eggs usually starts on the next day. Some strains of laying females can produce long, uninterrupted sequences of eggs. Male birds would have to attempt to mate with each female bird each day if, like most mammals, their sperm lost viability after a few hours at body temperature. Birds have therefore evolved a system of storing sperm within the female. The vagina of the female has sperm host tubules that store a small proportion of the sperm ejaculated by the male. Some of these sperm are then released from these tubules before each ovulation and reach the infundibulum at the same time as the ovum.

Sperm storage by poultry can be highly efficient. Some turkey hens have been found to produce fertile eggs 60 days after their last insemination, although a period of 28 days is more typical for the species. The length of fertile egg production after the last insemination varies between 6 and 12 days for geese, quail, ducks and chickens.

The Ovary

Female poultry have two gonads in their bodies. The left-hand gonad grows in a developing female embryo up to hatching and ultimately becomes the functional ovary. The gonad on the right of the bird regresses before the chick hatches, although it remains as a rudiment throughout the life of the female bird.

All female poultry have a juvenile period before egg production starts. When a female is in the first half of its juvenile period, its ovary weighs less than 0.1% of its body weight. However, the ovary of a sexually mature bird may be up to 3% of its body weight. Many thousands of tiny ova are connected to the ovary in an immature female bird. Individual ova are contained within small sacs, called follicles. A progressive series of follicles begins to increase in size as the bird approaches sexual maturity. An individual follicle changes from grey to grey-white as it grows, and then changes to pale yellow and finally bright yellow as yolk material is accumulated in the few days before ovulation. Yolk accumulation rapidly changes the weight of the follicles. An individual follicle will increase its weight 10-fold in the last 8 days before ovulation.

A mature bird that is in the middle of producing a long sequence of eggs has a series of follicles of decreasing weight. The heaviest one will most probably be released next and so on. Most immature follicles will never develop into full sized ova. Some may start to develop and increase in size and unaccountably regress. The yolk material from these ova is reabsorbed and used for other developing ova.

Occasionally the ovary of the female bird may be destroyed, most usually by disease. The hen will not ovulate any more nor will she lay eggs. The vestigial gonad on the right-hand side of the body may be stimulated to develop in these circumstances. This right gonad then has the characteristics of a male testis and may start to produce male hormones. This bird will then develop male characteristics such as a distinctive feather structure and comb shape and it also may crow.

Egg Formation Tract (Fig. 4.1)

A mature ovum has the exact chemical composition of the yolk in a laid egg. The rest of the egg components are added in an egg formation tract, also called the oviduct. The egg formation tract leads from the ovary, which is close to the left kidney in the body cavity, to the cloaca of the bird. It is approximately 65 cm long and weighs about 75 g in a mature laying hen. The size of the tract depends on the maturity and laying condition of the bird. The egg formation tract in a 10-week-old juvenile

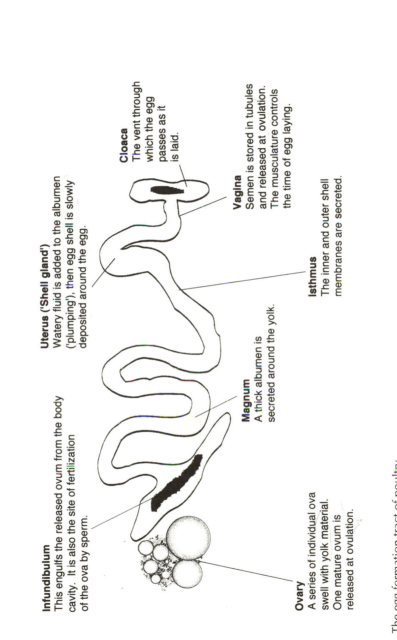

Infundibulum
This engulfs the released ovum from the body cavity. It is also the site of fertilization of the ova by sperm.

Uterus ('Shell gland')
Watery fluid is added to the albumen ('plumping'), then egg shell is slowly deposited around the egg.

Cloaca
The vent through which the egg passes as it is laid.

Vagina
Semen is stored in tubules and released at ovulation. The musculature controls the time of egg laying.

Magnum
A thick albumen is secreted around the yolk.

Isthmus
The inner and outer shell membranes are secreted.

Ovary
A series of individual ova swell with yolk material. One mature ovum is released at ovulation.

Fig. 4.1. The egg formation tract of poultry.

pullet is one hundredth of the weight of the tract in a mature laying hen. The egg formation tract begins to regress when a hen stops laying eggs. A regressed tract may shrink to a 10th of its previous weight.

Ovulation occurs when the follicle ruptures and releases the ovum into the body cavity. This is controlled by a release of luteinizing hormone (LH) from the pituitary gland 3–4 h beforehand. The LH surge is linked to a stimulatory release of progesterone from the ovary. A fully formed egg derived from the ovulation on the previous day is usually laid at almost the same time.

The released ovum is engulfed by the egg formation tract. Engulfing of the ovum is not always successful and the ovum can be irretrievably lost into the body cavity. This is called internal laying. Internal laying is more common shortly after a bird reaches sexual maturity. Up to 40% of all the ova shed in the first 2 weeks after some birds reach sexual maturity may be lost into the body cavity.

A correctly engulfed ovum is not modified further and becomes the yolk of the freshly laid egg. All the other egg components are secreted as the yolk moves down the egg formation tract. The tract has five distinct areas:

1. The **infundibulum** is the structure that engulfs the ovum. A defective or diseased infundibulum may cause a high incidence of internal laying. The infundibulum is also the site of fertilization of an ovum by the male sperm.
2. The **magnum** deposits nearly all the egg white protein, but little of the water that is bound with it, in the formed egg. This is the longest part of the oviduct. The forming egg moves slowly through by rhythmic contractions of the wall of the magnum.
3. The inner and outer shell membranes are formed in the **isthmus**. A small amount of albumen may also be produced.
4. Shell deposition occurs in the **uterus**, also called the shell gland. The steady progress of the egg down the oviduct is delayed when the egg reaches this pouch in the oviduct. Eggs may stay in the uterus for over 20 h. Glands in the uterus produce a watery fluid for the first few hours after the part-formed egg arrives. The fluid hydrates the albumen protein within the shell membranes and 'plumps' out the egg to its characteristic shape. The epithelial cells of the uterus secrete calcium salts. The calcium deposits initially bond onto the shell membranes. The changing structure of the deposited egg shell is controlled actively by a calcium carrier protein and by changes in pH. A pigment, porphyrin, may also be secreted in the final few hours of shell formation. Porphyrin produces the brown coloration on some egg shells.

5. The **vagina** takes no part in .the formation of any egg parts. A sphincter marks the border between the uterus and the vagina. The vagina is very short and its walls have a powerful musculature. Relaxation of the muscles allows the egg to leave the uterus and it is almost immediately laid through the cloaca. Vaginal muscles are under the voluntary control of the bird. The bird can therefore delay egg laying for a few hours if the situation is unfavourable.

EGG LAYING PATTERNS

Timing of Egg Laying (Fig. 4.2)

Most poultry take more than 24 h and less than 30 h to form their eggs. The time of egg laying is not just a random event and there are characteristic patterns for the different species. Domestic fowl lay most of their eggs in the morning and quail lay most of their eggs in the late afternoon. Turkeys lay their eggs in the late morning to early afternoon whereas ducks lay their eggs shortly after first light in the morning.

Ovulation is controlled by the timing of the pre-ovulatory release of luteinizing hormone (LH). The LH release only occurs in a restricted period each day, called the 'open period'. This results in the typical egg laying patterns of the flock. Open periods of domestic fowl are approximately 8–10 h and their timing is controlled by the biological clock of the bird. Light is a major stimulus that controls the biological clock. The main peak of egg laying activity in a flock of laying hens occurs about 16.5 h after the start of the previous dark period. Laying birds kept in darkness still have evident 'biological clocks' that respond to other environmental variables.

Ovulation gets nearer the end of the open period each day because egg formation usually takes longer than 24 h. No ovulation occurs when the end of the open period is reached. This pause in egg formation ends the sequence of daily egg laying. The next ovulation occurs at the start of the open period on the subsequent day.

Figure 4.3 gives a theoretical model of egg laying patterns of three laying hens that have different egg formation times. Egg formation times vary between individuals within a flock of birds. The egg formation time of an individual determines the number of eggs that it can lay within the 8 h open period for egg laying. The egg formation times of individuals within a flock are the main characteristics that determine the number of eggs laid by a flock. Large eggs take longer to form and egg weights and egg formation times increase as birds get older.

This model of the timing of egg laying simplifies the egg laying patterns of birds. In practice, sequences of eggs from an individual bird

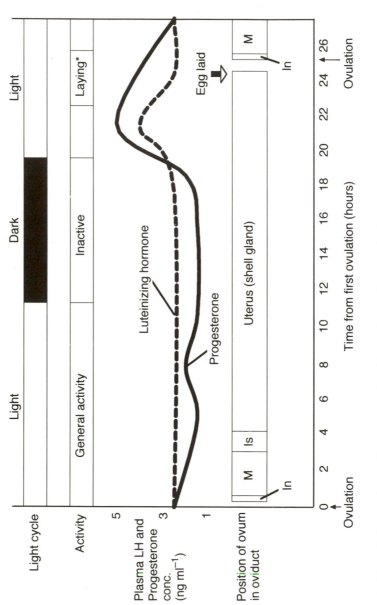

Fig. 4.2. Diurnal changes in egg formation and activity in laying domestic fowl. Position of ovum in oviduct: In = infundibulum, M = magnum, Is = isthmus.

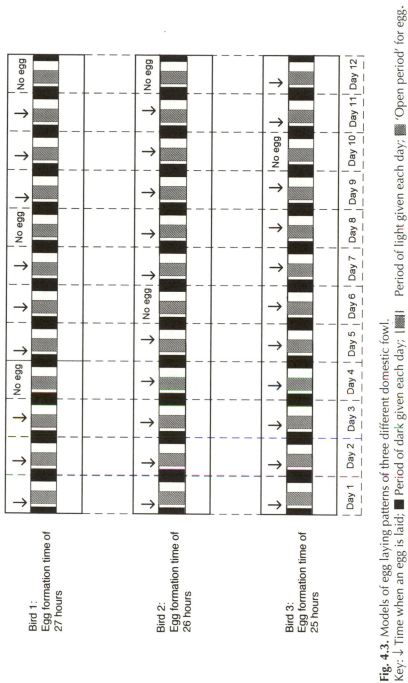

Fig. 4.3. Models of egg laying patterns of three different domestic fowl.
Key: ↓ Time when an egg is laid; ■ Period of dark given each day; |▨| Period of light given each day; ▨ 'Open period' for egg.

may include egg formation times that vary from less than 24 h up to 30 h. However, the model helps in understanding the effect of egg formation time on egg production.

Birds kept in natural lighting have periods of dim light close to sunrise and sunset. Laying hens kept in light-controlled buildings are usually given abrupt changes from bright light to complete darkness. Both lighting systems give equivalent effects on the photoperiodic responses of the birds. A period of dim light may be beneficial to encourage roosting by birds in some housing systems with controlled lighting. The dim light is interpreted by the birds as part of the daylight hours. Lighting programmes with no darkness may also be used. A daily cycle of bright light followed by dim light is interpreted by the birds as equivalent to light and dark periods respectively.

Photoperiodic Responses

The seasonal changes in the egg production of poultry are well known. Day lengths and changes in day lengths are used by birds as the primary factor that synchronizes their seasonal breeding patterns. Increasing day lengths stimulate egg production in birds.

A simple model of the photoperiodic response of poultry is shown in Fig. 4.4. Plasma LH concentrations are low when day lengths are below the critical day length for that species. Egg production will therefore stop, or not begin, if day length is less than this. Increases in day lengths above the critical day length stimulate increased LH release. Egg laying is therefore stimulated. However, a point is reached when any further increase in day length does not give any further increase in plasma LH. This is called the saturation day length. The bird only responds to day length changes when they are greater than the critical day length and less than the saturation day length. The intermediate range is called the marginal day lengths.

Critical and saturation day lengths are about 10 h and 14 h respectively for domesticated poultry. Prolonged exposure to long days will shift the critical and saturation day lengths upwards. This is called relative photorefractoriness. Turkeys are different. The turkey's critical and saturation day lengths are fixed. This is called absolute photorefractoriness.

Understanding the effects of photorefractoriness is important in managing commercial flocks of egg laying birds. Correct lighting patterns are essential to maximize the number of saleable eggs produced by a flock. The lighting strategies depend on whether the species exhibit absolute or relative photorefractoriness. Day lengths can be controlled

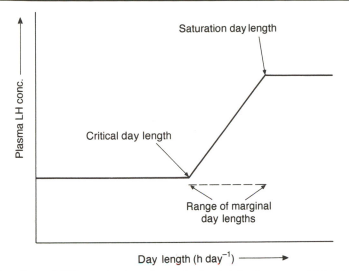

Fig. 4.4. The effect of different day lengths on the luteinizing hormone (LH) output of poultry.

either by housing the flocks in light-proof buildings or by supplementing the natural daylight.

Absolute photorefractoriness

Juvenile turkeys will never reach full sexual maturity unless their critical day length is reached. Turkey breeder hens can therefore be held on short days until they are old enough to lay eggs that are large enough for successful incubation.

Turkey breeder flocks can be quickly brought into production by abruptly increasing the day length to their saturation day length. Eggs will start to be produced after about 10 days. Fourteen hours per day is a satisfactory day length. There is no advantage to increasing day lengths beyond this because the saturation day length is fixed.

Relative photorefractoriness

Domestic fowl are an example of a species with relative photorefractoriness. Juvenile birds will eventually reach sexual maturity even if they are kept on short days. Juvenile breeder flocks cannot be held out of production by lighting alone. A feed restriction is usually given along with short day lengths to delay sexual maturity. Chickens that are reared at their critical or marginal day lengths reach sexual maturity earlier if increased day lengths are then given.

The saturation day lengths of the birds start to increase once egg production begins. The highest saturation day length that occurs after a prolonged exposure to long days is 17 h day^{-1}. A slow increase in day lengths above the original saturation day length maintains high plasma LH levels and gives the most egg numbers over the laying period.

Egg-strain domestic fowl are often reared on constant days at approximately 8 h day^{-1}. Day lengths are increased when the birds are old enough for egg production. Day lengths are then increased by between 0.5 and 1 h day^{-1} each week. The increases are stopped once the day length reaches 17 h day^{-1}.

EGG PRODUCTION CHARACTERISTICS

Egg Numbers

An individual laying bird can produce long sequences of eggs on succeeding days. She will then have a pause day with no egg. Further sequences of eggs will then follow. Individuals within a large flock will have different pause days and so the flock will have a smooth egg production curve that is typical for that class of bird.

The egg number curve can be divided into three main periods (Fig. 4.5):

1. Period 1 is the time between when the first eggs are laid until the time when nearly all the birds are laying continuously. The individuals within a flock are usually given the same environmental and nutritional regimens during rearing so they reach sexual maturity at very similar ages. Period 1 therefore is usually quite short.
2. Period 2 is the main laying period. This may last for various lengths of time that depend on the species and strain of bird and the environment in which the flock are kept. The continuous decline in the numbers of eggs laid during this period is mostly due to a lengthening egg formation time. However, there is also a slow and continuous reduction in the rate of egg yolk deposition as the birds get older. The birds are not able to form ova of the correct size quickly enough to allow long sequences of eggs to be formed.
3. The number of shed ova declines rapidly in period 3. The incidence of internal laying also increases sharply. Broody behaviour, moulting or

changes in nutrient intakes along with changes in body composition are the major causes of the end of egg laying (see later sections).

Figure 4.5 shows the typical characteristics of egg production curves. Seasonal variations in feed supply, feed quality, day length, temperature and disease challenges can change the shape of these egg production curves. However, flocks of poultry kept in controlled environment buildings are often able to achieve these production characteristics in most parts of the world.

Egg Weight

Different species of poultry, and strains within species, all have their characteristic egg weights (Table 4.1). The average egg weight of a flock of poultry increases as the birds get older (Fig. 4.6). The age of the bird, and not the previous number of eggs laid, is the main factor that determines egg weight. A flock that has a delayed sexual maturity will

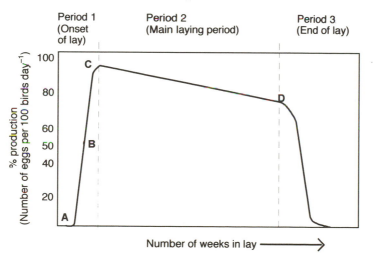

Fig. 4.5. Characteristics of an egg production curve of a flock of laying hens. Notes about the curve: The age that a flock reaches sexual maturity and lays the first egg is a characteristic of the species and strain but it is also influenced by management factors. The time of first egg (A) is a measure of age of sexual maturity but individuals can vary, so age at 50% production (B) is a more reliable measure. The flock rapidly reach a peak of egg production (C), thereafter egg numbers decline each week. The rate of decline is usually constant and the slope depends on the species, strain and management and health factors. Decreasing daylengths, a reduced food supply or a sharp increase in the number of broody birds may then cause egg laying to rapidly reduce (D).

Table 4.1. Egg production characteristics of poultry.

Species	Age at first egg (weeks)	Egg numbers at peak (% production)	Maximum egg numbers in one year	Mean egg weight (g)
Domestic fowl – egg laying strain	21	93	310	60
Domestic fowl – meat strain	24	80	160	65
Turkey	32	74	110	85
Guinea fowl	28	70	160	40
Quail	5	86	240	10
Common duck	21	92	270	65
Goose	38	45	40	130

lay eggs with the same weight as those from a flock with a normal age of sexual maturity.

Environment and diet can change the egg weights of a flock. High temperatures cause egg weight to be reduced. Lighting programmes that reduce the rate of ovulation of birds increase their egg weights (see later section). Low dietary protein or low linoleic acid concentrations may also reduce egg weights.

Egg Mass Output

Egg mass output is calculated from the number of eggs produced by a flock multiplied by the average weight of their eggs. Egg mass output rises to a peak shortly after a flock has reached peak egg number

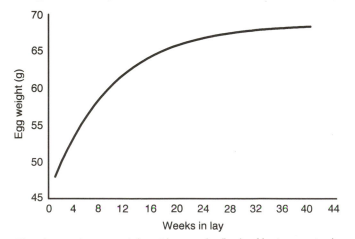

Fig. 4.6. The change in egg weight with age of a flock of laying-strain domestic fowl.

production and after that it decreases steadily until egg production ceases (period 3).

Egg Composition

There are only small changes in the composition of eggs over the laying period. The proportion of yolk is low in the first few eggs laid, but it quickly increases and after that remains constant.

The proportion of shell in the egg slowly decreases over the laying period. Shell strength remains high, or deteriorates only slowly, over a long part of the laying period. The shell strength then rapidly deteriorates.

Age at Sexual Maturity

The age that a bird first starts to lay eggs is a characteristic of the species and, to a lesser extent, a characteristic of strains within species. However, the age that an individual flock reaches sexual maturity can be advanced or delayed by changes in day length during rearing.

Three photoperiodic factors influence a bird's age at sexual maturity: longer day lengths decrease the age at sexual maturity if constant day lengths are given during the whole rearing period. An increase in day length during rearing reduces the age at sexual maturity. Moreover, the age when the changes in day length occur is important. Increases in day length when the birds are close to sexual maturity are more stimulatory than increases earlier in the juvenile period.

PREDICTION OF EGG PRODUCTION CHARACTERISTICS

Equations that predict the changes in egg numbers and egg weights over the laying period and the age of sexual maturity have been developed.

Egg Numbers

The equation given below gives an approximate estimate of the egg numbers produced by the birds in a flock until they reach the end of their laying period.

$$y = 100 \times ((1/(1 + (a \times b^x))) - (c \times x) + d \qquad (4.1)$$

where x = number of weeks from the first egg laid by the flock; y = the numbers of egg laid per day per 100 birds (percentage production); a and

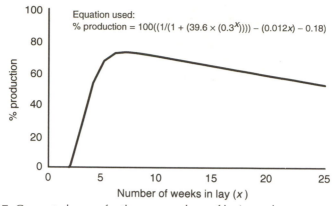

Fig. 4.7. Generated curve for the egg numbers of laying turkeys.

b = constants that describe the increase in egg numbers from the start of lay up to the peak of production; c = a constant that describes the rate of decline in percentage production from peak; and d = a constant that describes the percent production at peak. Some values for the constants that describe the typical egg number curves of different species are given in Table 4.2 and an example curve is shown in Fig. 4.7.

Egg Weight

The egg weights of laying birds increase quickly and then slowly move towards a constant weight as they get older. Equation 4.2 describes this asymptotic relationship between age and egg weight.

Table 4.2. Coefficients for the prediction of egg numbers.

Species	Coefficient a	Coefficient b	Coefficient c	Coefficient d
Domestic fowl – egg laying strain	39.6	0.3	0.0035	-0.03
Domestic fowl – meat strain	39.6	0.3	0.0120	-0.08
Turkey	39.6	0.3	0.0120	-0.18
Guinea fowl	39.6	0.3	0.0070	-0.22
Quail	39.6	0.3	0.0040	-0.11
Common duck	39.6	0.3	0.0040	-0.06
Goose	2.8	0.6	0.0300	-0.25

Prediction equation: Egg numbers (% production) = $100 \times ((1/(1 + (a \times b^x))) + (c \times x) + d)$.
x = number of weeks in lay.
Coefficients a and b for geese are different because they are the only species that are rarely kept in controlled environment housing during rearing. Flocks of geese therefore tend to be more variable in their age at sexual maturity. Their rate of reaching peak production is therefore slower.

Table 4.3. Coefficients for the prediction of egg weight.

Species	Coefficient *a*	Coefficient *b*
Domestic fowl – egg laying strain	62.0	18
Domestic fowl – meat strain	68.7	23
Turkey	90.8	13
Guinea fowl	41.6	8
Quail	11.2	4
Common duck	66.5	19
Goose	190.0	60

Prediction equation: Egg weight (*g*) = *a* − (*b* × 0.9ˣ).
x = number of weeks in lay.

$$y = a - (b \times 0.9^x) \tag{4.2}$$

where x = age in weeks; y = egg weight in grams; a = a constant that describes the maximum egg weight; and b = a constant that describes the rate of increase in egg weights. Values for the two coefficients a and b are given in Table 4.3 and coefficient r is usually 0.9 for all poultry species.

Egg gradings

Several different egg marketing schemes operate around the world but many sell eggs and pay producers according to the weight of individual eggs. Prices often depend on the consumer demand for a particular egg size and the price may not be closely related to the total weight of eggs produced. The proportions of eggs within a particular size grade can be an economically important factor in the egg weight characteristics of a flock.

Although the mean egg weight of a flock is a useful indicator of probable egg gradings, the variability of individual egg weights around the mean is also important. For example, the United States and Canadian egg marketing regulations have a size grade called 'extra large' that includes eggs weighing 63.8 g or more. A flock of birds that has a mean egg weight of 61.7 g with an typical variability (for example a standard deviation of 6.35 g) would be expected to have 37% of eggs in the extra large size grade. Another flock that has exactly the same mean egg weight of 61.7 g but with less variability in egg weight (standard deviation of 3.50 g) would have only 27% of the eggs in the extra large size grade.

The proportion of eggs above or below a specified egg weight can be calculated from the equation.

$$z = \frac{(\text{Specified weight of egg} - \text{Mean egg weight of flock})}{\text{Standard deviation of individual egg weights}} \tag{4.3}$$

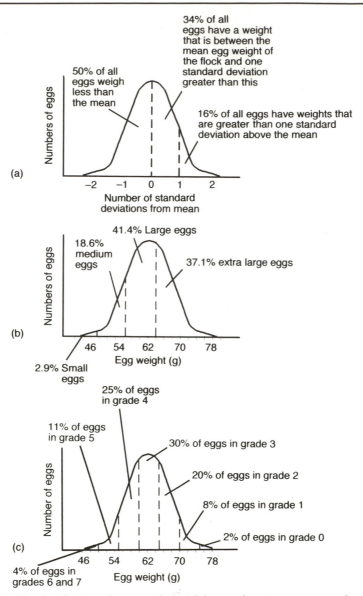

Fig. 4.8. The distribution of egg weights and the numbers in egg size grades.
(a)   The weights of individual eggs are symmetrically distributed about the mean
     egg weight of flock. Half of all eggs are lighter than the flock mean and half are
     heavier than the flock mean. Standard deviations indicate the amount of
     variation around the mean. One standard deviation greater or less than the
     mean will include 34% of all egg weights.
       The normal standard deviation can be calculated if the difference between
     any given egg weight and the mean egg weight is divided by the standard
     deviation. The proportion of eggs that have an egg weight in this range can be
     read from normal standard deviation tables.

The value, z, is called a standardized normal deviate and the proportion of eggs greater or less than the specified weight can be taken from standardized normal distribution statistical tables (see abbreviated data in Table 4.4 and Fig. 4.8). Standard deviations of individual egg weights can be measured by individually weighing a sample of eggs from the flock. They often have a range of 6–12% of the mean egg weight of the flock.

Age at Sexual Maturity

The age that a flock of poultry reaches sexual maturity depends upon its species and strain, but changing day lengths during rearing also have a large effect. The effects of photoperiods have been best quantified in layer strain domestic fowl. Changing day lengths are more important when the birds are close to sexual maturity but, as a general rule, each one hour increase in day length during rearing advances the age of sexual maturity by one and a half days.

The length of an unchanging daylength that is given each day also affects the age of reaching sexual maturity. This is a non-linear relationship as shown in equation 4.4 below.

$$\text{Age at sexual maturity} = \text{General average for strain} - 1.61P + 0.0006P^2 + 0.001918P^3 \tag{4.4}$$

where P = constant day length (photoperiod) used.

(b) An example flock has a mean egg weight of 61.7 g with a standard deviation of 6.35 g.

  The graph indicates the proportions of eggs from the example flock that would be expected in the United States egg size grades. The US size grades are:

  Extra large eggs: Egg weights of 63.8 g or greater

  Large eggs: Egg weights between 56.7 g and 63.7 g

  Medium eggs: Egg weights between 49.6 g and 56.6 g

  Small eggs: Egg weights of 49.5 g or less.

(c) The graph indicates the proportions of eggs in the European Union egg size grades from the same example flock with a mean egg weight of 61.7 g and a standard deviation of 63.5 g. The EU size grades are:

  Grade 0: Egg weights of 75.0 g or greater.

  Grade 1: Egg weights between 70.0 g and 74.9 g

  Grade 2: Egg weights between 65.0 g and 69.9 g

  Grade 3: Egg weights between 60.0 g and 64.9 g

  Grade 4: Egg weights between 55.0 g and 59.9 g

  Grade 5: Egg weights between 50.0 g and 54.9 g

  Grade 6: Egg weights between 45.0 g and 49.9 g

  Grade 7: Egg weights of 44.9 g or less.

FACTORS THAT STOP EGG LAYING

Eventually a flock will stop laying eggs altogether. This may be due to a single factor or a combination of factors.

Photorefractoriness

Changes in day length are used to synchronize seasonal breeding patterns. Decreasing day lengths result in decreased luteinizing hormone secretion by birds and the ceasing of egg production. Flocks kept in controlled environment houses can be given lighting programmes that never decrease day lengths and may give regular small increases in photoperiod. However, this stimulatory effect diminishes over time and the birds eventually stop laying eggs. The loss of photostimulation in these birds is called photorefractoriness (see earlier section).

Broodiness

The time during which a female bird incubates her eggs and broods her young is called the 'broody' period. In natural conditions, broodiness begins when the bird completes the clutch size that is characteristic of

Table 4.4. Normal standard deviation table for predicting the proportions of eggs below a specified weight.

Standardized normal deviate (z)	Proportion of egg weights less than the specified weight
0.0	0.500
0.2	0.579
0.4	0.655
0.6	0.726
0.8	0.788
1.0	0.841
1.2	0.885
1.4	0.919
1.6	0.945
1.8	0.964
2.0	0.977
2.2	0.986
2.4	0.992
2.6	0.995
2.8	0.997
3.0	0.999

More detailed tables can be found in most statistics textbooks.

her species. Broody birds become protective of their chosen nest sites and they have a characteristic hiss or cluck. Most important, they stop laying eggs and their ovaries regress. Feathers on the abdomen of the bird are lost and they eventually form a bare brood patch when they start incubating their eggs.

Broodiness usually occurs after a period of high egg production by the individual bird. The levels of circulating prolactin increase during this time and probably reach a threshold level that starts overt broody behaviour. There are also large changes in circulating LH and progesterone levels and other gonadotrophic hormones at this time.

A broody period typically lasts for the same time as the incubation period for that species. The bird will not lay any eggs during this time. Heavy strains of turkeys are particularly prone to broodiness although it occurs to some extent in all strains and species of laying birds. Domestication and selection have reduced the high incidence of broodiness from egg-strain chickens but it is still a problem in heavier meat-strain chickens.

Large-scale poultry production systems do not need broody behaviour in poultry. Fertile eggs are invariably removed from the nest boxes or cages and incubated and reared separately from the parent birds. A high incidence of broodiness can result in a large drop in the number of eggs produced by a flock. Controlling the birds' environment, identifying the early stages of broodiness in individual birds and treating the early stages of broodiness are major tasks in managing breeder flocks.

Birds kept on littered, solid floors are more likely to become broody than birds kept in wire-floored cages. These systems provide more places that suit the birds' preferences for nest sites. A bird that shows early signs of broodiness may be returned to egg production quicker by some management intervention. Early broody behaviour is usually spotted by a stockworker checking for birds that are restless and lingering around a nest site. These birds are then usually isolated in an unfamiliar environment, for example a broody coop, until the broody behaviour subsides. A broody coop is a well-lit, wire bottomed cage in which the bird has easy access to food and water. The hens are kept in the broody coop for about 5 days and they will probably resume laying after about 25 days. Injected gonadotrophic hormones reduce the physical characteristics of broodiness but the birds do not come back in to lay any faster compared to the stockmanship methods.

Moulting

Birds kept in egg production for a long period eventually have a rapid deterioration in their egg production and shell strength. Egg production then soon ceases altogether. The oviduct and ovary regress and most of the bird's feathers are shed. The bird's feather papillae are stimulated to produce new feathers that push the old ones out. The process of the ending of egg production and feather loss is called moulting. Moulting is a result of a complex interaction between the gonadotrophic hormones. Other hormones, such as thyroxine and prolactin, are also known to interact with the gonadotrophic hormones. The physiological mechanisms that initiate a moult are poorly understood.

Moulting provides the bird with a rest from egg production and allows time for tissue regeneration, particularly in the oviduct. Birds kept outdoors often synchronize the start of a moult with a period of low food supply. The bird will then remain out of egg production until the food supply improves or there is an increase in day length. Moulting can be induced in flocks in controlled environment houses by reducing food intakes and daylengths.

Egg production declines once a moult is begun and ceases completely after about 10 days. The oviduct regresses and its weight reduces to about one-tenth of its previous weight. The ovary also regresses and reduces to one-twentieth of its previous weight. Weight loss of the ovary and the oviduct may account for approximately 25% of a laying hen's body weight loss during a moult. Liver weights and body fat stores are also reduced.

Feather loss starts about 15 days after the start of the moult. Although birds lose their feathers throughout their first laying period, these feathers are not replaced. Progesterone levels are lowered during a moult and this appears to stimulate the feather follicles to start feather regeneration. The feather loss is caused by new feathers pushing out the old ones. Laying hens lose between 1–6 primary wing feathers and up to 13 secondary feathers. New feather growth continues for up to 100 days after a moult is begun. Low nutrient intakes in the early laying period after a moult can stunt feather growth and reduce egg output.

Moulted birds will start egg production again before feather growth is fully completed. The changes in the circulating gonadotrophic hormones at the start of the second laying period generally mirror the changes at the start of the first laying period. Once a moulted flock starts its second period of lay, its egg production quickly rises to a peak that is 8–12 eggs per 100 birds per day greater than would have been expected at this age (Fig. 4.9). The weekly rate of decline after the second peak is usually very similar to that achieved in the first laying period. The increase in egg weight as a flock ages is not altered by a moult. However, shell strength and internal egg quality, for example Haugh units, are improved.

Individuals within a flock will naturally moult at different ages. A procedure that synchronizes the moult of all the birds in the flock has

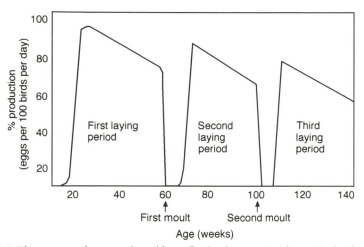

Fig. 4.9. The egg numbers produced by a flock of egg-strain domestic fowl moulted twice (at 60 and 100 weeks of age).

many practical advantages. Consequently, several induced moulting techniques have been developed. The procedures usually involve quantitative and/or qualitative feed restriction along with reduced photoperiods and/or reduced light intensity (e.g. see Table 4.5).

Different moulting procedures need to be given for different species, for strains within species and for different management systems. The length of the egg laying pause during an induced moult is influenced by the length of the feed restriction period. Increasing the length of the egg pause (up to 4 weeks in laying chickens) gives an increase in egg numbers and an improvement in egg quality characteristics in the

Table 4.5. An example of a moulting procedure suitable for brown egg laying strain chickens.

Time (days)	Feed type	Feed allowed (g per bird day⁻¹)	Photoperiod (h light day⁻¹)	Light intensity (lux at darkest spot)
Prior to moulting	Complete layers feed	ad lib	≥14 h	≥ 10 lux
Start of induced mould				
Days 0–8	Whole grain cereal	15 g	8 h	2 lux
Days 9–18	Whole grain cereal	30 g	6 h	2 lux
Days 18–28	Whole grain cereal	60 g	6 h	2 lux
Days 29 onwards	Complete layers feed	ad lib	10 h with a weekly increase to ≥ 14 h	≥ 10 lux

Egg production would cease completely around day 8. The second laying period would begin about 50 days after the start of the induced moulting programme.

second laying period. Long periods of feed restriction achieve a uniform body fat loss throughout the flock and give a complete regression of the oviduct. These factors may be important in optimizing egg production and egg quality in the second laying period.

Other dietary methods have been used to induce moults in laying hens: ad lib access to low calcium diets, low sodium diets and high zinc diets all cause a moult. Although all these methods give an egg laying pause, they may not always give a complete moult. For example, low calcium and low sodium diets give less feather loss compared to quantitative feed restriction methods.

LIGHTING PROGRAMMES

Simple Light–Dark Programmes

If birds are kept in controlled environment houses then artificial light needs to be provided. A single light and dark period each 24 h is the simplest programme and most practical lighting programmes are based on this system. Poultry species that have only relative photorefractoriness, for example domestic fowl, benefit from small increases in the light period throughout their laying period. An effective lighting programme must link the lighting given during rearing to the lighting programme during lay. Often birds are reared on relatively short constant day lengths, for example 8 h each day, and then given weekly increases in day length once they are placed in the laying accommodation. Small weekly increases in day length will be continued, up to a maximum of 17 h per day, throughout a large part of the laying period. Figure 4.10 gives an example of a practical lighting programme for domestic fowl kept in controlled environment buildings. Birds kept outdoors may have supplementary lighting provided in their night-time accommodation at times of the year when there would otherwise be decreasing daylengths.

Intermittent Light Cycles

Profound changes in lighting programmes can be used with flocks kept in light controlled buildings. Intermittent light cycles reduce electricity use by giving less light and, more important, may improve the efficiency of food utilization by a flock. Two types of intermittent cycles can be used.

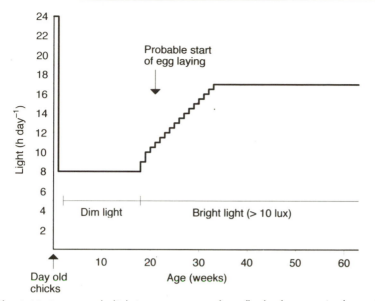

Fig. 4.10. An example lighting programme for a flock of egg-strain domestic fowl.

Intermittent cycles using 24 h day lengths

Poultry interpret the length of a light period from the time light first occurs until the start of the longest dark period. Periods of darkness can then be added into the birds' perceived period of light without affecting its photoperiodic effect. An example of this type of intermittent light cycle is given in Fig. 4.11.

These types of cycles maintain the same circadian rhythms in the birds as expected with a conventional lighting programme. Laying activity is also very similar. There is a reduction in bird activity during the dark periods that occur within the birds' perceived photoperiod. These periods of inactivity reduce the energy requirements of the birds by a small amount. The efficiency of feed utilization may therefore be improved. Second, the reduced time for feeding activity may reduce the daily feed intakes of the birds. Birds given these lighting programmes often have less body fat. A 15% reduction in mortality over the laying period is typical in birds given these intermittent cycles compared to conventional cycles. The reduced body fat composition may be a major factor that improves bird survival.

Short repeating intermittent cycles

Light:dark cycles that repeat every 4, 6 or 8 h are also successfully used for laying domestic fowl. For example, a 6 h cycle gives about 1.5 h of light followed by the remaining 4.5 h of darkness. Four 6 h cycles will be completed each 24 h. These types of intermittent cycles also use less

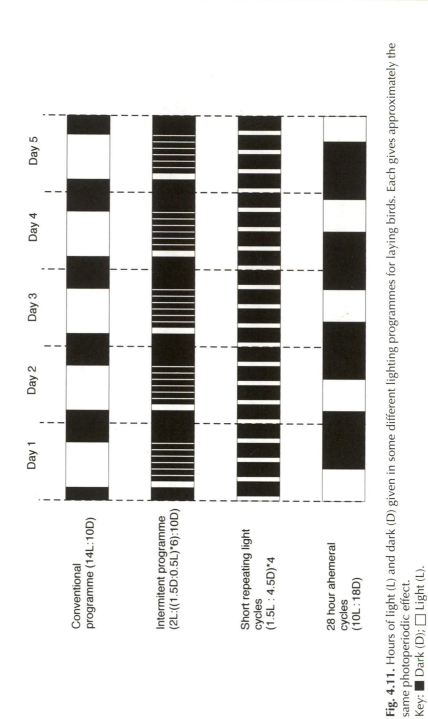

Fig. 4.11. Hours of light (L) and dark (D) given in some different lighting programmes for laying birds. Each gives approximately the same photoperiodic effect.

Key: ■ Dark (D); □ Light (L).

light and give a small improvement in feed utilization efficiency. Egg weights may be increased by these light cycles; however, egg mass output is almost invariably unchanged.

Ahemeral Cycles

Light:dark cycles that do not add up to a natural 24 h day are called ahemeral cycles. Birds given ahemeral cycles ranging from 21–30 h will synchronize their egg laying patterns to these cycles. These entrained flocks will have marked peaks in the time when egg laying occurs. Most of the eggs will be laid within an 8 h open period. Flocks of birds given ahemeral periods outside 21–30 h will not become entrained and they have egg laying patterns that are similar to birds in continuous light.

Birds that are given ahemeral cycles less than 24 h will only have a short sequence of eggs before the end of their open period is reached and a 'pause day' is necessary. The high number of 'pause days' will result in a lower number of eggs produced.

Long ahemeral cycles have more advantages in practical poultry production systems. Six 28 h ahemeral cycles fit exactly into one week and so are relatively easy to arrange with conventional electrical switching equipment. Nearly all the birds within a flock of young domestic fowl hens could probably form an egg in less than 28 h. When these birds are given a 28 h lighting programme, they will form an egg each light cycle and no pause days will occur. These birds will produce six eggs a week because there are only six 28 h light cycles each week. This rate of egg production is equivalent to 85.7 eggs per 100 birds per 24 h period. Flocks that would peak at greater than 90 eggs per 100 birds per 24 h in 24 h light cycles would therefore produce fewer eggs when given 28 h ahemeral cycles (see Fig. 4.12).

The longer time between successive ovulations allows for more yolk material to be deposited. The weight of the yolk is increased and more albumen is deposited to surround it. Over the whole week of egg production the total egg mass produced by the flock is close to, but still less than, the egg mass output of an equivalent young flock given 24 h light cycles.

The egg spends most of the extra time in the egg formation tract in the uterus so extra time is available for shell deposition. Eggs from hens given long ahemeral cycles therefore have stronger shells. Birds take longer to form each individual egg as they age and a point may be reached where most birds take longer than 28 h hours to form each egg. If these birds were given 28 h hour cycles they would have fewer pause days and would produce a greater numbers of eggs than with 24 h cycles.

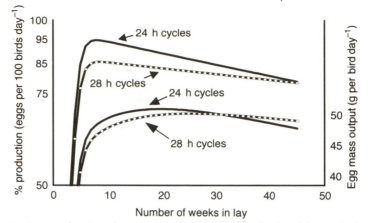

Fig. 4.12. Egg production characteristics of two similar flocks of domestic fowl given either conventional 24 h light cycles or 28 h ahemeral light cycles.

Hens given 28 h ahemeral light cycles have very different production curves compared to flocks given 24 h light cycles. Only a few days are needed to entrain a flock to a different light cycle. A flock of birds can be moved on and off a 28 h ahemeral lighting cycle from a conventional 24 h cycle. Their production characteristics would quickly move on to the egg production curve that is characteristic for that light cycle length.

The use of long ahemeral cycles in commercial egg production can give a few practical problems and production differences for practical egg production systems.

1. A lighting programme that does not repeat every 24 h is not always convenient for the stockworker. Periods of darkness within the lighting programme will coincide with the daylight working hours. Bird inspection, equipment maintenance, feeding and egg collection are then impossible. However, a programme of bright and dim light, instead of light and dark, can be used to entrain the birds to the ahemeral cycles. The intensity of the bright light must be 30 times greater than the dim light to successfully entrain birds to a 28 h ahemeral cycle. The dim light can be made bright enough to allow work within the laying house at all times.

2. The peak of egg laying activity in flocks of hens is typically 13–16 h after the start of the last dark period so most domestic fowl eggs are laid in daylight hours in the mid to late morning. Long ahemeral light cycles cause an interaction between the synchronizing effects of the onset of darkness and the endogenous 24 h circadian rhythm of the birds. This affects the timing of the LH release that determines egg laying activity. Flocks given 28 h ahemeral cycles lay most eggs 6–9 h after the start of the dark period. The peak of egg laying comes in the dark. This is not a problem when the birds are kept in cages, but higher numbers of floor eggs would be expected if a loose-housing system was used.

The timing of the peak of egg laying of a laying hen flock given any length of long ahemeral cycles can be predicted from the equation given below.

Time of peak of egg laying $= 64.62 - 2.161C + 0.268S$ (4.5)
(h after onset of darkness)

where C = ahemeral cycle length (h); and S = length of dark period used in the ahemeral cycles (h).

3. A flock of birds can be completely entrained to a long ahemeral lighting cycle. Their patterns of egg laying, feeding and general activity will follow this light programme exactly. However, an underlying, endogenous 24 h circadian rhythm continues. This affects the photoperiodic response of the bird when a change is made to a different light cycle. For example, a flock of birds previously given a 24 h cycle of 14 h light : 10 h dark is moved onto a 28 h cycle of 14 h light : 14 h dark. The light periods do not coincide with the endogenous 24 h circadian rhythms of the birds and light is given outside the 'day time' periods of the birds' circadian rhythms. The birds interpret this as an increase in day length. The light hours need to be reduced to give no effective change in photoperiodic stimulation. A reduction in light hours each cycle by the same time as the increase in cycle length is necessary. For example, a flock that had previously been given a 24 h cycle of 14 h light : 10 h dark would need to be given a 28 h cycle of 10 h light : 18 h dark to result in no change in photoperiodic stimulation. This rule of setting up ahemeral lighting programmes is especially important if it is intended to shuttle between a number of light cycles over the productive life of a laying flock.

FURTHER READING

The Reproductive Tract and Egg Formation

Bell, D.J. and Freeman, B.M. (eds) (1971) *Physiology and Biochemistry of the Domestic Fowl. Volume 3*. Academic Press, London.

Etches, R.J. (1996) *Reproduction in Poultry*. CAB International, Wallingford, UK.

Freeman, B.M. and Lake, P.E. (eds) (1972) *Egg Formation and Production*. Poultry Science Symposium No. 8. British Poultry Science Ltd, Edinburgh.

Solomon, S.E. (1991) *Egg and Eggshell Quality*. Wolfe Publishing Ltd, London.

Egg Laying Patterns

Fraps, R.M. (1961) Ovulation in the domestic fowl. In: Villee, C.A. (ed.) *Control of Ovulation*. Pergamon Press, New York, pp. 133–162.

Phillips, J.G., Butler, P.J. and Sharp, P.J. (1985) *Physiological Strategies in Avian Biology*. Blackie, Glasgow.

Sharp, P.J. (1980) Female reproduction. In: Epple, A. and Stetson, M.H. (eds) *Avian Endocrinology*. Academic Press, New York, pp. 435–454

Silver, R. (1986) Circadian and interval timing mechanisms in the ovulatory cycle of the hen. *Poultry Science* 65, 2355–2362.

Wilson, W.O. (1964) Photocontrol of oviposition in gallinaceous birds. *Annals of the New York Academy of Sciences* 117, 194–202.

Egg Production Curves and Prediction Equations

Abs, M. (1976) Physiology of juvenile development and sexual maturation in non-passerine birds – a review. *Archiv für Geflugelkunde* 5, 153–167.

Adams, C.J. and Bell, D.D. (1980) Predicting poultry egg production. *Poultry Science* 59, 937–938.

Factors that Stop Egg Production

Decuypere, E. and Verheyen, G. (1986) Physiological basis of induced moulting and tissue regeneration in fowls. *World's Poultry Science Journal* 42, 56–68.

Guemene, D. and Williams, J.B. (1992) Changes in endocrinological parameters and production performances in turkey hens (*Meleagris gallopavo*) submitted to different broody management programs. *Theriogenology* 38, 1115–1129.

Sharp, P.J., Dunn, I.C. and Cerolini, S. (1992) Neuroendocrine control of reduced persistence of egg-laying in domestic hens: evidence for the development of photorefractoriness. *Journal of Reproduction and Fertility* 94, 221–235.

Wolford, J.H. (1984) Induced moulting in laying fowls. *World's Poultry Science Journal* 40, 66–73.

Lighting Programmes

Bhatti, B.M. and Morris, T.R. (1988) Model for the prediction of mean time of oviposition for hens kept in different light and dark cycles. *British Poultry Science* 29, 205–213.

Morris, T.R. (1973) The effects of ahemeral light and dark cycles on egg production in the fowl. *Poultry Science* 52, 423–445.

Morris, T.R. (1988) Use of intermittent lighting to save feed and improve egg quality in laying flocks. *Proceedings of the 18th World Poultry Congress*, pp. 161–164.

Shanawany, M.M. (1992) Response of layers to ahemeral light cycles incorporating age at application and changes in effective photoperiod. *World's Poultry Science Journal* 48, 156–164.

Male Reproduction 5

THE MALE REPRODUCTIVE TRACT

The purpose of the male reproductive tract is to produce viable sperm and transfer them into the female's vagina. Semen production and maturation are more rapid in poultry than in mammals. Newly formed sperm may appear in the ejaculate after only 13 days. Sperm are viable when they are ejaculated into the vagina and they do not need to be further matured as in some mammals. They can remain viable within the male reproductive tract for over a month before being ejaculated. Sperm can also be stored in the female's vagina for many weeks without losing viability.

Sperm Production

Male birds have two testes between the bottom of the lungs and the top of the kidneys. They therefore produce viable semen at deep body temperatures of 43°C. The testes weigh about 1% of the mature weight of most poultry species (Fig. 5.1).

Seminiferous tubules occupy the greater part of the testes. The epithelium of the tubules has several layers and all stages of sperm production occur in this area. Germ cells are present in the tubules. A series of spermatogonial divisions result in primary spermatocytes being formed from there. The primary spermatocytes undergo a meiotic cell division to form daughter cells, called secondary spermatocytes. Secondary spermatocytes divide to form spermatids that then undergo a metamorphosis to form sperm. The spermatids and sperm are attached to Sertoli cells that give them a supply of nutrients. Sertoli cells control the release of the sperm into the lumen of the tubules and control the flow of seminal fluid through the tubules. The seminal fluid and sperm then pass in to the ductus deferens.

Each ductus deferens is a convoluted duct that shares a common connective sheath with the ureter. The ductus deferens has four main purposes:

1. Transport of semen from the testes to the cloaca.
2. Maturation of sperm. The sperm only gain their fertilizing ability once they have entered the ductus deferens.
3. Storage of semen before ejaculation. The storage ability is small. One half to two-thirds of the contents of the ductus deferens are expelled at each ejaculation.
4. Unejaculated semen is broken down and resorbed if the semen has been stored for too long. Sperm can remain fertile for up to 35 days when stored in the ductus deferens.

Each ductus deferens leads to the cloaca. Within the cloacae of domestic fowls, turkeys, guinea fowl and quail are two small round folds with a small central white body. This phallus becomes engorged with a transparent fluid at copulation and it protrudes from the cloaca by 1–2 mm. The small erect phallus has a central groove through which semen passes. Some transparent fluid may leak out of the phallus and mix with the semen. Most males within a breeder flock will copulate at least daily

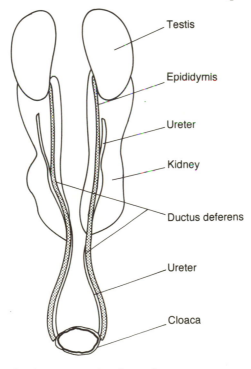

Fig. 5.1. The reproductive organs of male poultry.

but the frequency depends on their pecking order with other males in the pen.

Male geese and ducks have prominent penises that are spirally twisted when erect. Semen flows down a groove that goes down the length of the penis. The cloaca of female ducks and geese does not evert at mating and the male's penis acts as an intromittent organ.

SEMEN

Characteristics

Poultry sperm are small compared to mammals. They are filamentous, about 100 μm long, with a small cylindrical head that contains DNA. Sperm are contained within an opaque white seminal fluid that is 96% water and has a high concentration of sodium, potassium and glutamate compared to mammalian semen. During copulation, a variable amount of transparent fluid filters out of the engorged phallus and mixes with the semen. Transparent fluid has a similar composition to lymph and is alkaline. Variation in the amount of transparent fluid that mixes into the semen can give a pH that varies between 7.0 and 7.6. Transparent fluid improves the fertilizing ability of freshly ejaculated semen but the sperms lose viability quicker during storage.

Some species, particularly quail, also release a frothy liquid with the semen. The froth is produced from glands at the base of the phallus and has a very similar composition to transparent fluid. The clustering of sperm is reduced and sperm motility is increased in the presence of this froth. Guinea fowl do not produce any transparent fluid or froth with their semen.

Production

Male and female birds within one flock will reach sexual maturity at very similar ages. The photoperiodic changes that affect the sexual maturity of female birds affect males similarly (see Chapter 4). Semen volume and semen quality reach their maximum levels about 3 weeks after they reach sexual maturity. If the male birds are removed from the females and given increasing day lengths before the female birds they will reach sexual maturity earlier. A high quality semen is then available to inseminate the females when they lay their first eggs.

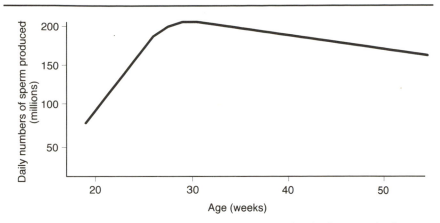

Fig. 5.2. The daily spermatozoa production of a strain of male domestic fowl (adapted from deReviers and Williams, 1984).

Total sperm production is the volume of semen multiplied by its concentration of sperm. Sperm output rises to a peak shortly after male birds reach sexual maturity and after that there is a continuous decline (Fig. 5.2).

Male domestic fowl produce 0.5–1.0 ml of semen per ejaculation and turkeys produce only about 0.2 ml of semen. The number of matings each day affects the volume of semen per ejaculate and the concentration of sperm within the semen. Three ejaculations a week can produce high quality semen from an individual for a short time. High semen production is obtained when males are kept in low stress situations such as individual pens, with no feed restrictions and with relatively low environmental temperatures.

Natural mating systems keep male and female birds together in large pens. The semen quality of the males can only be altered by controlling the proportion of males to females in the flock. A high proportion of males reduces the number of matings by each individual so a greater number of sperm are inseminated at each mating.

ARTIFICIAL INSEMINATION

Advantages of Artificial Insemination

Artificial insemination involves keeping male and female birds separately. Semen is collected from the males and is then injected into the vagina of females. Artificial insemination needs a highly skilled labour input and the technique is costly to use in practical poultry systems. Nevertheless, it has many advantages.

1. Males of the highest genetic worth can be used more efficiently in breeding programmes. For example, a domestic fowl male can provide enough semen to artificially inseminate 40 females. A maximum ratio of one male to 10 females is used in natural mating systems. Sire and dam records in pedigree flocks are easier to keep, and more accurate, if artificial insemination is used.

2. Artificial insemination can ensure a high fertilization rate in a flock. Males mate more frequently with females in the middle of their pecking order. Those females at the top or bottom of this order may not mate frequently enough to always have enough viable sperm in their oviduct. Artificial insemination ensures that each breeding female gets an optimum amount of male semen. The physical size of the poultry strain may also affect the fertilization rate in the flock. The rapid genetic progress in altering the weight and shape of turkeys has already created these problems. Adult male turkeys are much heavier than the breeding females. The weight difference and the protruding breast of the male result in many unsuccessful mating attempts. Artificial insemination is the only practical way of keeping high fertility in heavy turkey flocks.

3. Artificial insemination allows breeding birds to be kept in small group sizes. They can be kept in small cages without having to have a male in each cage. Laying guinea fowl are difficult to manage in loose-housing systems because male birds are very aggressive. Guinea fowl breeding systems often keep the females in small groups in cages and the males in separate cages. Artificial insemination is used to fertilize the females. Hatching duck egg programmes have also used artificial insemination methods for these reasons.

Practical Methods of Artificial Insemination (Fig. 5.3)

Semen collection

Similar methods are used to stimulate ejaculation in all species of domestic fowl. The back of the bird is massaged or stroked close to its tail while the operator also applies a slight finger pressure around the base of the tail. The phallus should then become erect within the cloaca. Pressure is then applied around the cloaca and the tail is flattened towards the back of the bird causing the phallus to protrude from the cloaca. The operator's thumb is then pressed on the bird's abdomen directly beneath its vent. This causes semen to be released from the ductus deferens almost immediately and the operator gently squeezes, or 'milks,' the semen from the swollen papillae at the base of the phallus.

An aspirator can be used to suck up the semen into a tube. The quality of semen produced by a single bird at each ejaculation varies

with each milking. The semen should appear milky-white and dis-
coloured semen often indicates excessive transparent fluid or excreta
contamination. The aspirator not only collects semen quickly but also
high quality semen can be 'picked-off' and the poor quality semen can
be discarded. Waterfowl have long coiled penises and it is more difficult
to milk the semen. A wide mouthed glass tube that tapers towards the
bottom is sometimes used for these birds instead of an aspirator. The
penis is inserted into the tube that collects the whole volume of the
ejaculate.

The correct management of male birds is important to maximize
their yield of high quality semen. Birds should be handled gently but
quickly. They become used to individuals and a change of the semen
collector may reduce semen volumes for a few days. The birds must be
handled frequently before semen collection starts so that they become

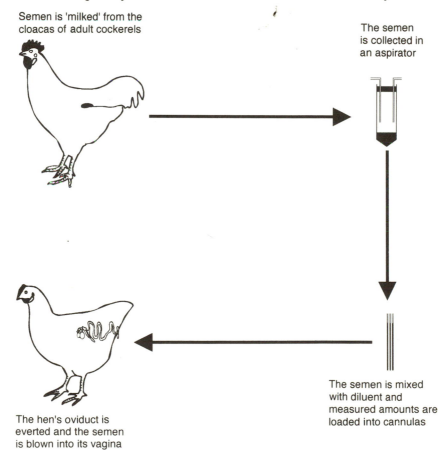

Semen is 'milked' from the cloacas of adult cockerels

The semen is collected in an aspirator

The semen is mixed with diluent and measured amounts are loaded into cannulas

The hen's oviduct is everted and the semen is blown into its vagina

Fig. 5.3. A practical method of artificial insemination of domestic fowl.

accustomed to the procedure. They must also be handled regularly throughout their period of semen production although their semen may not be required.

Any contamination of the semen can cause a loss of sperm motility. All collection apparatus must be scrupulously clean. The sperm may suffer cell damage if the collection vessel is too cold. Sperm survival is greatest at 10°C, but the collected semen must be cooled slowly to this temperature as rapid cooling may result in cell damage and loss of viability of the sperm.

Semen dilution and storage

There are major differences between the semen storage characteristics of poultry and mammals. Mammalian semen can be mixed with an extender and stored for up to 4 days with little loss of fertilizing ability. The viability of poultry semen is seriously reduced after 48 h storage. Turkey semen loses fertility particularly quickly and a high loss of fertility can occur if turkey semen is held for more than 30 minutes.

Storage of poultry semen is more successful if an extender is used. Semen extenders are also called diluents because a second function for them is to reduce the concentration of spermatozoa. For example, only 0.01 ml of undiluted semen needs to be inseminated into a turkey hen to give around 100 million sperm. These small volumes are difficult to measure and inseminate efficiently and a much higher semen volume is often recommended because of this. The addition of a diluent allows larger semen volumes to be inseminated without oversupplying sperm. Dilution rates of 1 part semen : 1 part diluent up to a 1 : 3 ratio are most suitable.

Extenders and diluents need to have the same features.

1. They must have an osmotic pressure that is slightly hypertonic compared to the semen. This is around 360 mosmol kg^{-1}.
2. They should provide a balance of minerals for the sperm cells.
3. There should be an energy source for sperm metabolism. Either glucose or fructose is usually provided.
4. They should provide buffering capacity to maintain pH of around 7.0–7.4.
5. They should attempt to protect sperm from bacterial organisms.

Semen diluent formulations can be quite simple. For example, semen can be held for up to 15 minutes in a 1% solution of sodium chloride. Equal parts of a 3% sodium citrate solution and egg yolk has also been used as a short term diluent. More complicated mixtures need to be used for longer-term storage above 0°C or for freezing.

There are also differences between poultry and mammals in the success of long-term storage of semen. Glycerol, or other cryogenic agents, are added to mammalian semen. The sperm can then be frozen in liquid nitrogen at -196°C for many years. Freezing of poultry semen is possible but the fertility of the thawed semen is often very poor. Glycerol is an effective agent to protect sperm cells against damage at freezing but the same compound damages poultry sperm cells when they are thawed. Glycerol must be quickly removed on thawing or no fertile sperm will remain. Frozen poultry semen may be used to conserve exceptional genetic material in breeding programmes but a high loss of sperm would be expected.

Insemination of females

The hen's oviduct must be exposed from the cloaca before she is inseminated. A person holds the bird, head down, between their legs or across their chest. The oviduct is everted by pressing the tail towards the back and by gently pressing the bird's abdomen with the other hand. The left-hand side of the abdomen is pressed because only the left oviduct is functional in the female bird.

Male semen is put in thin plastic straws, called cannulas. Each cannula should contain the correct volume of semen to give 100 million sperm. The cannula is inserted into the everted oviduct deep enough to deposit sperm close to the sperm storage glands in the vagina. Best depths range from 20 mm for smaller domestic fowl, guinea fowl and ducks to 60 mm for heavy strain turkeys and geese.

The semen is blown out of the cannula into the vagina. Semen can leak back out of the oviduct when the empty cannula is withdrawn, so the semen needs to be blown out of the cannula just before the oviduct is allowed to revert. Semen loss can be reduced by this correct method of insemination. The oviducts of geese and ducks cannot be everted easily. The inseminator has to find the oviduct in the left-hand side of the cloaca with a finger and the cannula is then passed along the side of the finger and into the bird's vagina.

FURTHER READING

The Male Reproductive Tract

Lake, P.E. (1984) The male in reproduction. In: Freeman, B.M. (ed.) *Physiology and Biochemistry of the Domestic Fowl. Volume 5*. Academic Press, London, pp. 381–406.

Lake, P.E. and Furr, B.J.A. (1971) The endocrine testis in reproduction. In: Bell, D.J. and Freeman, B.M. (eds) *Physiology and Biochemistry of the Domestic Fowl. Volume 3*. Academic Press, London, pp. 1469–1488.

Semen

de Reviers, M. and Williams J.B. (1984) Testis development and production of spermatozoa in the cockerel (*Gallus domesticus*). In: Cunningham, F.J., Lake, P.E. and Henritt, D. (eds) *Reproductive Biology of Poultry*. Poultry Science Symposium No. 17. British Poultry Science Ltd, Harlow, Essex, pp. 183–202.

Fujihara, N. (1992) Accessory reproductive fluids and organs in male domestic birds. *World's Poultry Science Journal* 48, 39–56.

Lake, P.E. (1989) Recent research in male reproduction. In: Nixey, C. and Grey, T.C. (eds) *Recent Advances in Turkey Science*. Poultry Science Symposium No. 21. Butterworths, London, pp. 55–68.

Lake, P.E. and El Jack, M.H. (1966) The origin and composition of fowl semen. In: Horton-Smith, C. and Amoroso, E.C. (eds) *Physiology of the Domestic Fowl*. Oliver and Boyd, Edinburgh, pp. 44–51.

Lake, P.E. and Wishart, G.J. (1984) Comparative physiology of turkey and fowl semen. In: Cunningham, F.J., Lake, P.E. and Hewitt, D. (eds) *Reproductive Biology of Poultry*. Poultry Science Symposium No. 17. British Poultry Science Ltd, Harlow, Essex, UK.

Artificial Insemination   ·

Bootwalla, S.M. and Miles, R.D. (1992) Development of diluents for domestic fowl semen. *World's Poultry Science Journal* 48, 121–128.

Clayton, G.A., Lake, P.E., Nixey, C., Jones, D.R., Charles, D.R., Hopkins, J.R., Binstead, J.A. and Pickett, R. (1985) Artificial insemination. In: *Turkey Production: Breeding and Husbandry*. Reference Book No. 242, MAFF. HMSO, London.

Cooper, D.M. (1982) Artificial insemination. In: Gordon, R.F. and Jordan, F.T.W. (eds) *Poultry Diseases*. 2nd edn. Bailliere Tindall, London, pp. 365–373.

Lake, P.E. and Stewart, J.M. (1978) *Artificial Insemination in Poultry*. Bulletin 213, MAFF. HMSO, London.

Sexton, T.J. (1984) Breeding by artificial insemination. In: Cunningham, F.J., Lake, P.E. and Hewitt, D. (eds) *Reproductive Biology of Poultry*. British Poultry Science Ltd, Harlow, Essex, UK.

Wambeke, F.van and van Wambeke, F. (1990) Semen preservation above 0°C in fowls: interest and current limits for industrial purposes. *Colloques de l'INRA*, 54, 177–184.

Incubation 6

Poultry have a remarkable ability to get high proportions of their fertile eggs to develop into healthy hatchlings. Up to 90% of the fertile eggs incubated by broody hens may hatch and modern artificial incubators achieve similar hatchability rates. Each of the commercial poultry species have characteristic incubation times (Table 6.1) although these times can be slightly changed by different incubation temperatures. Eggs stored before incubation also hatch slightly later than eggs incubated immediately.

DEVELOPMENT OF THE EMBRYO (Fig. 6.1)

A newly laid, fertile hatching egg contains approximately 56% albumen, 32% yolk and 12% shell. There is a 12% reduction in egg weight by the time the chick hatches. This is due to water loss. The weight of the shell hardly changes during incubation but almost all the rest of the egg contents are converted into a hatched chick (Fig. 6.2).

Table 6.1. Typical egg incubation times of avian species.

Species	Days
Domestic fowl	21
Turkey	28
Common duck	28
Muscovy duck	30
Goose	30
Guinea fowl	25
Quail	16

Morphology

The first stages of embryo development are started before the egg is laid. The ovum is fertilized shortly after ovulation so cell division continues during its movement through the egg formation tract. A small disc of cells, the blastoderm, that lies between the yolk and its vitelline membrane forms the embryo and this is visible in the laid egg. On incubation, the cells within the blastoderm align themselves towards the outside of the disc and then form a line down the centre. This is called the primitive streak.

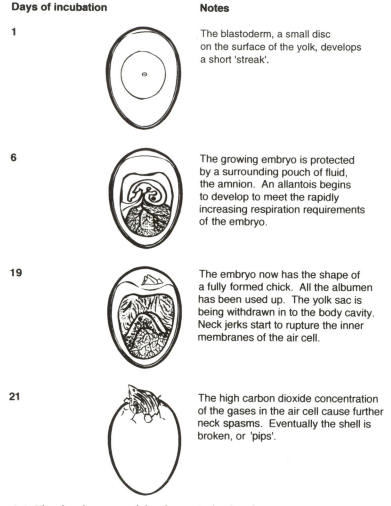

Days of incubation

Notes

1

The blastoderm, a small disc on the surface of the yolk, develops a short 'streak'.

6

The growing embryo is protected by a surrounding pouch of fluid, the amnion. An allantois begins to develop to meet the rapidly increasing respiration requirements of the embryo.

19

The embryo now has the shape of a fully formed chick. All the albumen has been used up. The yolk sac is being withdrawn in to the body cavity. Neck jerks start to rupture the inner membranes of the air cell.

21

The high carbon dioxide concentration of the gases in the air cell cause further neck spasms. Eventually the shell is broken, or 'pips'.

Fig. 6.1. The development of the domestic fowl embryo.

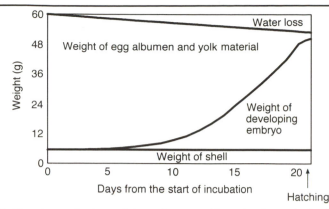

Fig. 6.2. The change in the weight and composition of a domestic fowl hatching egg during incubation.

Distinct buds are evident along the body of the chicken embryo 3 days after incubation starts. These buds will ultimately develop into limbs. A heart begins to function after 3 days although it is outside the 'body' at first. The digestive tract becomes completely closed on the 5th day and lungs, with their characteristic air sacs, are evident on day 6. Dense areas of feathers begin to form at 8 days. The calcification of the skeleton begins about day 10 and is complete by the 15th day. Beaks and claws are complete by day 16.

Nutrient Supply

The albumen, yolk and shell contain all the nutrients required by the developing embryo. At first, the embryo is small enough for it to obtain nutrients by diffusion. However as the embryo grows, its body detaches from the egg contents and becomes suspended in a fluid-filled sac, the amnion. An umbilicus extends into a yolk sac where blood is produced and a network of blood vessels is developed. Blood moves from the heart and circulates in the yolk sac to quickly get the rich supply of nutrients back to the embryo. The yolk sac, containing any unused yolk material, is drawn into the abdomen of the embryo before hatching. The remaining yolk is gradually used for food in the first few hours after hatching.

Excretion of Waste Material

The allantois, an impermeable sac attached to the end of the digestive tract, is evident after four days of incubation. Toxic waste products of metabolism, mainly urates, are excreted in solution through the kidneys and collected in the allantois. About 6 ml of fluid will accumulate by day 13 of incubation. The allantois becomes semipermeable around day 14 and then water is reabsorbed from the allantois back to the developing embryo until practically none remains by the 20th day. The allantois, and its contents of solid urate deposits, is left with the egg shell at hatching.

Respiration

The egg shell contains thousands of microscopic pores that allow gas exchange with the surrounding atmosphere (see Chapter 2). A fertile chicken egg absorbs 6 litres of oxygen and gives off 4.5 litres of carbon dioxide and 11 litres of water vapour during the 21 days of its incubation. The embryo forms a membrane, the chorioallantois, to aid the exchange of gases. The chorioallantois has a rich supply of capillaries and is attached to the shell membranes. It grows to cover the whole egg shell by around day 14. The chorioallantois is comparable to the placenta of mammals.

Oxygen enters the egg by diffusion and the rate of supply is sufficient until the embryo starts the rigours of hatching. The volume of the water lost during incubation is replaced in the egg by air that is absorbed through the shell into the air cell. The air cell thus swells during incubation to account for 15% of the volume of the egg. The first hatching movements of the chick rupture the air cell and there the hatchling takes its first breaths with its lungs. It then has a large and rapidly available source of oxygen to support its efforts to break through the egg shell.

HANDLING AND STORAGE OF HATCHING EGGS

Collection and Cleaning

Hatching eggs are laid either in nest boxes on litter material, or laid in cages on to wire floors. After laying, embryo development continues if ambient temperatures are above 20°C. The sooner the internal temperature of the egg can be brought below this critical temperature then embryo development can be arrested. Hatching eggs need to be collected two or three times a day if the house temperature is below 20°C, but more frequent collections are necessary at higher temperatures.

A major objective of the breeder farm management is to ensure that all eggs are laid in clean conditions, but some eggs inevitably become soiled even in well-managed systems. Dirty eggs should not be kept for incubation because their bacterial contamination could be spread to other eggs in the incubator. However, hatching eggs have a relatively high value and so there is often an economic advantage in taking the risk of incubating some of these eggs.

Reducing the bacterial contamination, particularly on dirty eggs, can effectively reduce the risk of large scale bacterial contamination of a batch of incubated eggs. This is achieved in three ways.

1. Dry cleaning. Any spots of contamination are removed from the egg shell by using a stiff brush, coarse sandpaper or wire wool.
2. Washing. Egg washing is an option if there is a lot of excreta on the shell. Proprietary egg washing machines usually use heated water that contains a detergent and sanitizer. The temperature of the washing water should be above 38°C or at least 12°C warmer than the eggs being washed. Most bactericidal sanitizers are not effective at a neutral pH. Acid solutions react with the calcium carbonate in the egg shell so an alkaline pH greater than 10 should be used.
3. Fumigation. Formaldehyde gas is a potent bactericide. Fumigation of newly laid hatching eggs kills shell bacteria before they penetrate the shell. Reacting formalin with potassium permanganate, or heating para-formaldehyde prills, produces formaldehyde gas. The fumigation must be conducted in a gas tight chamber with an air circulation fan or incorrect formaldehyde concentrations will result.

Storage

Poultry lay eggs over a number of days until they have a suitable clutch size before they start to incubate the eggs. Fertile eggs can therefore tolerate a wide range of short-term variations in their pre-incubation environmental conditions and still remain viable. However, there is a slow degeneration in the structure of the embryo as it is stored and the probability that an embryo will develop becomes less as storage continues (Fig. 6.3). Embryo growth ceases almost completely below 21°C, but the best embryo survival occurs when the eggs are stored at 11–18°C. However, the optimum storage temperature becomes lower as the storage time increases. Maintaining the optimum temperature is an important factor that affects the degeneration of the embryo because it slows the osmotic movement of water from the albumen to the yolk. High humidity reduces the loss of water from the egg but care must be taken to avoid condensation on the egg shells. Albumen pH continually rises as carbon dioxide is lost from the hatching egg during storage. Covering the egg with a low-permeability polythene film reduces the loss of carbon dioxide. Covering hatching eggs with these plastic films is beneficial, but only if the eggs are being stored for extended periods.

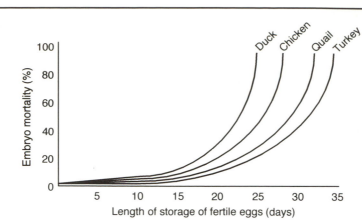

Fig. 6.3. The length of storage of hatching eggs and embryo mortality.

There is a small increase in embryo survival if hatching eggs are stored small-end upwards. An equally successful method is to store eggs broad-end up but turn them 90° each day if they have been stored for more than 7 days. These methods ensure that the yolk keeps a central position within the albumen and gives the correct orientation of the blastoderm.

PHYSICAL CONDITIONS NECESSARY FOR INCUBATION

A fertile egg is a self-contained life support system for the developing embryo. However, the hatching egg still depends on its environment, for heat, gas exchange and movement, to ensure that chick development continues (Fig. 6.4).

Temperature

The internal temperature of the egg is the most important physical factor that affects the development of the embryo (Fig. 6.5). The developing embryo hatches earlier if it is incubated at high temperatures up to a maximum of 39°C. However, a continuous 37.5°C gives the best rate of embryo survival. Incubator temperatures can be reduced by up to 2°C shortly before hatching because of the increased activity of the embryo.

Humidity

The relative humidity within an incubator affects the rate of evaporative water loss from the hatching egg. A relative humidity of 61% often gives the correct rate of loss of water, but other variable factors such as shell porosity, air movement and differences between strains can influence this. The most accurate method of adjusting humidity is to monitor the weight loss of the eggs. The weight of eggs at hatching should be 12% less than their weight at setting. A constant rate of weight loss usually occurs, so egg weights give a sensitive measure of the correct humidity setting in the incubator at any time during incubation.

Gas Concentration

The embryo depends on a supply of oxygen in the surrounding air. Fresh air contains about 21% oxygen at sea level. This is also the optimum oxygen concentration for developing embryos although they can tolerate a wide range of concentrations around this. Embryos are more susceptible to low oxygen concentrations and embryo survival is reduced below 15% oxygen. When eggs are incubated at high altitudes

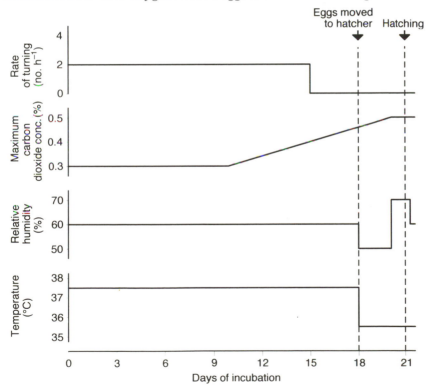

Fig. 6.4. A summary of the environmental conditions typically used in the large scale incubation of chicken eggs.

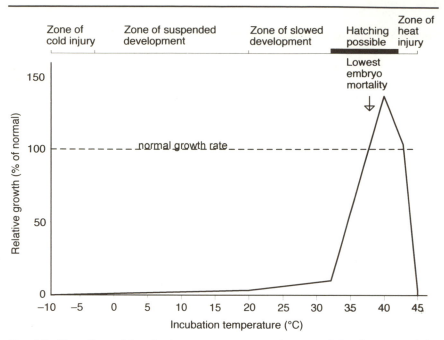

Fig. 6.5. The effect of incubation temperature on the rate of development of the domestic fowl embryo.

there can be a poor hatchability because of the low oxygen concentration in the air. If supplementary oxygen is added into these incubators, the hatchability of the eggs can be returned to normal.

High concentrations of carbon dioxide cause a reduced growth and a reduced viability of the developing embryos. Fresh air typically contains 0.03% CO_2. Concentrations of 0.4% significantly reduce the viability of developing embryos, although the embryos become more tolerant later in the incubation period. Inadequately ventilated incubators result in high CO_2 and low O_2 concentrations. The buildup of CO_2 often causes more hatchability problems than are caused by the lack of oxygen. Fan speeds that give eight changes of fresh air per hour generally give an adequate ventilation rate.

Orientation and Turning

Gravity affects the developing embryo. The first breath of air the developing embryo takes is from the air cell at the broad end of the egg. Without this breath the hatchling may lack enough oxygen to successfully break out of the egg shell. Eggs incubated with the broad end of the egg uppermost have the greatest proportion of embryos oriented with their heads towards the air cell. More than 90% of the hatchlings would be expected to be positioned this way.

Turning of eggs during incubation prevents the developing embryo adhering to the egg membranes and reduces the possibility of embryo mortality. Most large artificial incubators mechanically turn the eggs through a 90° arc up to four times each hour. Hatching chicken eggs are often turned for the first 18 days of incubation although turning between days 4 to 7 of incubation is the critical period. The turning must be gentle, especially in the first few days of incubation when the blood vessels are being organized. Shocks and jarring of hatching eggs may also separate the inner and outer shell membranes which then disrupts the positioning of the air cell.

COMMERCIAL INCUBATION METHODS

An artificial incubator is a closed box in which there is accurate control of temperature and control of ventilation and other environmental variables. There is an enormous range of possible sizes and sophistication of mechanized incubators. Low cost, small sized incubators often have just one small tray in the machine. A heater is regulated by a temperature sensor positioned at the same level as the eggs. Ventilation is regulated by adjusting the size of openings at the top or bottom of the machine. Convection currents within the machine control the rate of movement of air into or out of the incubator. A water container is placed into the machine so that water slowly evaporates. Altering the amount of evaporative surface area in the container gives an approximate control of the relative humidity. Turning of the eggs is done once or twice a day by hand.

Incubators that hold larger numbers of hatching eggs are often large cabinets. Eggs are placed in trays with the broad end upwards and the trays of eggs are placed on many levels throughout the machine. An even temperature distribution throughout the incubator is obtained by circulating air around the incubator with fans. Fans also control the ventilation rate into the incubator. Sensors monitor humidity and gas concentrations and are linked to devices that control their levels within the machine. Turning of the eggs is done mechanically. Some cabinet incubators may be big enough to walk in or even drive machinery into.

Single-stage incubators are designed to incubate eggs that are all at the same stage of development. The metabolic rate of the developing embryo increases rapidly towards the end of incubation and waste heat, up to 120 watts h^{-1} for each one thousand hatching eggs, is released. A simple system of cooling is needed to remove this relatively small

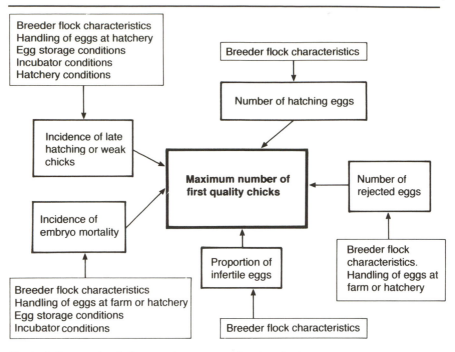

Fig. 6.6. Factors that influence the success of an incubation programme.

Table 6.2. A practical system of incubation of domestic fowl eggs.

Days from start of incubation	Task	Notes
−7 to −1	Egg storage	Eggs are collected each day to allow a large batch of eggs to be set at one time.
−1	Fumigate	Bacterial contamination of the shells is reduced.
0	Set eggs	Eggs are placed in the incubator.
6	'Candle' eggs	Inspection over a strong light allows clear eggs to be rejected. Clear eggs are either infertile or have had early embryo mortality.
18	Transfer eggs to hatcher	This coincides with a large change of environmental conditions for the hatching eggs.
21	Take away hatchlings	Each chick is inspected for viability and possibly the males and females are segregated.
22	Discard any remaining eggs	Late hatching chicks are not likely to be viable.

amount of waste heat. Multi-stage incubators incubate eggs at different stages of development in the same cabinet. The waste heat from the older embryos adds to the extra heat needed for maintaining the correct temperature.

Hatching eggs need different environmental conditions in the few days before the chick hatches. Many commercial hatcheries move the hatching eggs to different cabinets, called hatchers, a few days before hatching. This means that incubators and hatchers can be calibrated to their correct temperatures and humidities and then continuously run at these levels. The hatchers are usually sited away from the incubators to avoid cross contamination of bacteria between different batches of eggs.

A practical system of incubation of domestic fowl eggs is shown in Table 6.2. Figure 6.6 summarizes the factors that influence the success of an incubation programme.

FURTHER READING

Development of the Embryo

Deeming, D.C. and Ferguson, M.W.J. (eds) (1991) *Egg Incubation: its Effects on Embryonic Development in Birds and Reptiles.* Cambridge University Press, Cambridge.

Freeman, B.M. and Vince, M.A. (1974) *Development of the Avian Embryo.* Chapman and Hall, London.

Romanoff, A.L. (1960) *The Avian Embryo.* The Macmillan Company, New York.

Romanoff, A.L. (1972) *Pathogenesis of the Avian Embryo.* Wiley-Interscience, New York.

Physical Conditions for Incubation

Ar, A. and Rahn, H. (1980) Water in the avian egg: Overall budget of incubation. *American Zoologist* 20, 373–384.

Rahn, H., Ar, A. and Paganelli, C.V. (1979) How bird eggs breathe. *Scientific American* 240, 38–47.

Tullett, S.G. (1990) Science and the art of incubation. *Poultry Science* 69, 1–15.

Commercial Incubation Methods

MAFF (1977) Incubation and hatchery practice. MAFF Booklet No. 3060. HMSO, London.

Tullet, S.G. (ed.) (1991) *Avian Incubation*. Poultry Science Symposium No. 22. Butterworth-Heinemann, London.

Poultry Breeding and Genetic Improvement 7

The aim of breeders of commercial poultry stocks is to permanently change the phenotypic characteristics of the birds that improve their productive performance. Growth rate, egg production and egg quality, feed conversion efficiency, carcass conformation and disease resistance are examples of characteristics that poultry breeders aim to improve in their stocks. The phenotypic characteristics of individual poultry are generally a combination of genetic and environmental differences. However, genetic selection results in permanent changes in the poultry flocks and these changes will be transmitted to subsequent generations irrespective of any environmental differences.

Chromosomes are the carriers of genetic code within the bird and they occur in the nucleus of each cell in characteristic pairs. Domestic fowl, quail and guinea fowl have 39 pairs of chromosomes and ducks, geese and turkeys have 40 pairs. The two chromosomes in a pair are both very similar with one exception: the fifth largest pair of chromosomes in poultry may consist of two unequal chromosomes. This is a sex characteristic. Female birds have two unequal chromosomes, called ZW, whereas male birds have two similar chromosomes, ZZ. Female poultry are therefore the heterogametic sex, and males are the homogametic sex. This is in contrast to mammals where the male is the heterogametic sex.

Commercial poultry breeding programmes are nowadays conducted by fewer than twenty large companies throughout the world. The growth of meat birds and egg production in laying birds are the two most economically important characteristics. These two traits have negative genetic and phenotypic relationships, so selection of dual-purpose breeds is not worthwhile. Breeding companies either specialize in breeding meat-line birds or have two separate breeding programmes for meat and egg producing strains. These companies primarily use techniques of mass selection, index procedures and the development of specific sire and dam lines in their breeding programmes.

This chapter first describes the common methods used in commercial breeding programmes. There have been many studies of the phenotypic characteristics of poultry controlled by single gametes. The successes of these programmes have been limited, but a few useful monogenic traits are now used in commercial breeding programmes. Molecular techniques have the potential of contributing to breeding programmes in the future. Some early techniques of genetic manipulation in poultry are also described in this chapter.

COMMERCIAL METHODS OF POULTRY BREEDING (Fig. 7.1)

Egg-laying Poultry

Commercial breeding programmes exist for egg-laying strains of domestic fowl and ducks. The breeding programmes of these companies involve keeping up to 60 pure lines. Each of the pure lines will contain 200 to 300 females and 50 to 100 males. The birds are mostly kept in individual cages and the flock will be pedigree recorded; this means the parentage of all birds will be recorded so that the genetic relationships between individuals in a population are known.

The number of eggs laid in the course of one laying season has a relatively low heritability of 0.15 to 0.25. Also, this important production characteristic can only be measured in one sex. Therefore the best genetic gain per unit time is made when selections are based on a family index of performance. The selection index will include some of the following weighted breeding values.

1. Productive performance traits. Traits such as egg numbers, egg weight and feed efficiency are assessed. Records that use only the first part of the laying period may be used because the selected individuals can be then inseminated and used to provide offspring for the next generation. This avoids the need for sib testing in female birds and gives a much faster rate of genetic progress. Male birds are selected on the average performance of their female sibs (full sisters).
2. Egg quality traits. Desired levels of shell colour, shell strength and internal egg quality traits are maintained. The relative selection pressures used will depend on their economic importance in a particular market.
3. Disease resistance. Resistance to diseases such as Marek's Disease is heritable and selection for resistance can be beneficial though effective vaccines are available.

Different selection indices are used for most of the pure lines in a breeding programme to get different rates of genetic progress in important traits. Selection of several independent pure poultry lines enables poultry breeders to exploit the advantages of cross-breeding, called heterosis. The size of the improvement in productive performance due to heterosis is difficult to predict. Poultry breeding companies need to regularly examine test crosses of their pure and experimental lines to find superior crosses. A test crossing would probably involve

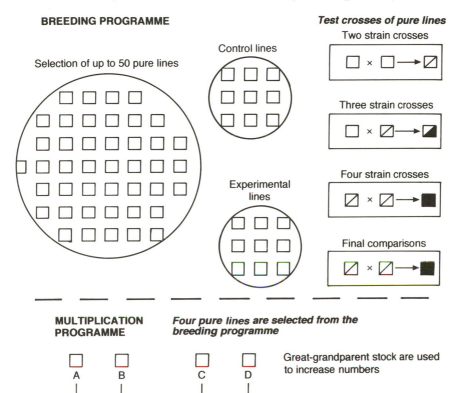

Fig. 7.1. A typical commercial poultry breeding programme.

evaluating around 200 birds. Some test crosses may be evaluated at more than one commercial site to see if there are any genotype × environment interactions. These interactions result in genotypes that have phenotypic characteristics that differ depending upon the growing site. Viable crosses are then tested on a bigger scale and compared against current commercial crosses. The breeding company will keep some control strains that have no selection pressure applied. This allows the company to evaluate the amount of genetic progress they are making in their selection programme.

Three or four pure lines will be selected for use in producing the commercial strain of egg-laying poultry. Parent female birds will be produced by cross-breeding two pure lines and parent male birds will be produced by cross-breeding two completely different pure lines. A single pure line may be used for the female parent birds. The crossing of the parent breeders benefits from heterosis and the commercial off-spring have good egg production characteristics.

Meat-type Poultry

Commercial breeding programmes exist for most species of meat-line poultry. The breeding companies keep a large number of pure lines. Growth rate was very important in the early days of breeding for meat-line domestic fowl (broilers). Many companies then used mostly mass selection techniques to select future parents from their pure lines. Large flocks of pure line birds would be reared with only a few of the most superior individuals being selected. The success of these breeding programmes has relied on the use of a simple genetic theory and selection for a primary trait (body weight) where genetic variation has not been depleted. Large breeding companies have the facilities and computer technology to be able to use large populations where they can use intense selection pressures. However, today's broiler industry re-quires genetic progress on several other bird characteristics as well as body weight. Characters such as feed conversion efficiency, carcass conformation, meat yields, chick viability, absence of leg or breast defects and good reproductive performance in parent birds are im-portant. Selection indices are therefore used to weight these breeding values. The need to assess characteristics where individuals have to be killed requires some amount of sib testing.

Breeding programmes now have fully or part-pedigreed pure lines and use progeny testing. Most meat-line domestic fowl and turkey strains do not thrive in cages, so a system of trap-nesting may be used to identify eggs from individual females. Artificial insemination tech-niques also simplify the identification of the male parent.

When a high selection pressure for growth rate is applied to a pure line then its reproductive performance may decline. A breeding programme needs to maintain a fairly high reproductive performance in the parent females otherwise the cost of producing the commercial chicks will be too high. Different selection criteria need to be applied to pure lines destined for use either as male or female lines. Selection of pure lines destined as male parents will give a high weighting for growth, whereas pure lines for female parents will have a greater selection pressure for reproductive success. Often, selection with female lines involves an early assessment of growth followed by using egg-laying records to compute a final selection index.

Many test crosses will be made between the pure lines to identify the superior crosses. Three or four pure lines will be selected for use in the multiplication programme and their numbers are increased to give grandparent stocks. Parent birds will be produced by cross-breeding two pure lines selected for male characteristics and two pure lines selected for female bird characteristics. The cross-bred male and female parent birds will be mated to derive commercial broiler chickens.

MONOGENIC TRAITS

Some phenotypic characteristics of poultry are primarily controlled by single gametes. The two gametes contributed by the two parents may contribute a factor that is dominant or recessive. Some very useful traits have been identified that are determined by a single dominant or recessive gene.

Feather Colour

The dominant gene for white feathers inhibits any black colouration. Black feather stubs spoil the appearance of poultry carcasses and so most meat-bird strains now carry this gene to give 'end-product' birds with white feathers.

Feather Sexing

A trait for early feather growth is linked to the sex chromosome. The dominant gene (K) for slow feather growth has been introduced into female broiler breeder lines. When these females are mated with a fast feathering homozygous male (k^+/k^+) their male offspring (K/k^+) are slow feathering and their daughters $(k^+/-)$ are fast feathering. The differences in feathering are seen when the chicks first hatch.

The ability to easily distinguish between males and females at hatching is an important commercial attribute. It is often more efficient to rear sexes separately. Feather sexing allows quick segregation at the hatchery. Vent sexing at hatching (distinguishing between males and females by the shape of papillae in their cloacas) is an alternative but it is slow and a highly specialized skill.

Dwarf Broiler Breeders

Some commercial breeding programmes use a recessive gene that reduces the mature body weight of broilers by 20–30%. The dwarfing gene (dw) is sex linked and situated close to the rate of feather growth allele. *any one of the DNA codings of the same gene*

The gene is introduced into the breeding programme by using a homozygous male (dw/dw) in the grandparent mating to produce the female parent stock. All the parent females will have the homozygous dwarfing gene ($dw/-$). When the parent males do not carry the dwarfing gene then both the females ($Dw^+/-$) and the heterozygous males (Dw^+/dw) will be a normal size.

Compared to normal parent breeders these female birds have lower feeding and housing costs whereas their reproductive performance is unaltered or even slightly improved. Unfortunately there is incomplete dominance of the Dw^+ gene and growth rates may be reduced by up to 3%. However dwarf broiler parent breeders are used when the reduction in the cost of producing hatching eggs outweighs the increased costs of production of the final broiler. This is usually when low slaughter weights are required.

The dwarfing gene is not available in turkeys and ducks. The gene is available in White Leghorn egg laying strains but it has not been used in commercial breeding programmes.

USE OF MOLECULAR TECHNIQUES

Traditional poultry breeding programmes can only introduce new genetic material into the population from mutations or by crossing with new populations containing desired traits. Molecular techniques allow new genetic material to be introduced artificially into the population.

Gene transfer in non-avian species typically involves identifying a segment of DNA of interest. A few hundred copies of the segment are then directly injected into the pronucleus of the embryo. Manipulation of the early poultry embryo has special problems because the fertilized zygote is inaccessible and the laid egg is a 24-hour-old embryo of about

60,000 cells. The transferred gene will only incorporate into all the genome (reproductive cells) of the bird if the genetic material is inserted at the one-celled stage.

Initial successes in gene transfer in poultry used RNA viruses. Infectious retroviruses are RNA viruses that penetrate the cell and

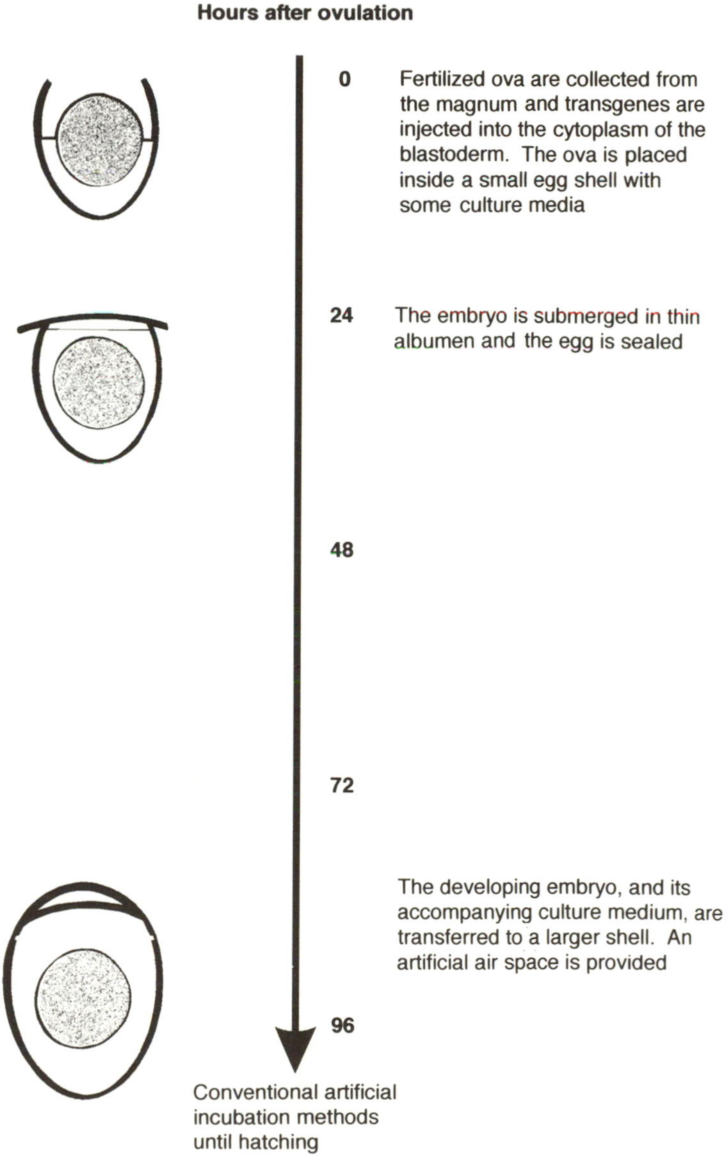

Hours after ovulation

0	Fertilized ova are collected from the magnum and transgenes are injected into the cytoplasm of the blastoderm. The ova is placed inside a small egg shell with some culture media
24	The embryo is submerged in thin albumen and the egg is sealed
48	
72	
	The developing embryo, and its accompanying culture medium, are transferred to a larger shell. An artificial air space is provided
96	

Conventional artificial incubation methods until hatching

Fig. 7.2. A method of direct transgene injection in poultry embryo (after Mather, 1994).

become integrated with the host cell DNA in a predictable orientation. Their main disadvantage is that they have little space available for gene insertion, their efficiency of transferring the foreign genes is low and they may give the birds a reduced immunity to other infectious retroviruses.

Recent gene transfer work has developed a method of directly injecting DNA into the pronucleus of the chicken embryo. Ova are collected soon after they enter the magnum after they have been fertilized in the infundibulum of a hen. Trans-genes are then injected into the cytoplasm of the pronucleus and the ovum is placed in a windowed egg shell that contains thin albumen and a culture medium (Fig. 7.2). The ovum, and its accompanying albumen and culture medium, are moved to a larger shell after four and a half days. The developing embryo is then left to develop until hatching at 21 days.

The production of novel or therapeutic proteins secreted by transgenic chickens into eggs may be an early commercial application of genetic manipulation of poultry. Products with medicinal uses are currently being produced in other transgenic farm animal species and the future production of compounds in eggs from transgenic poultry has the potential to be an efficient production method.

The early application of trans-genes in breeding programmes to improve the performance of commercial poultry will probably be to solve specific problems, such as conferring disease resistance. Little is yet known about the nature or location on the chromosome of the genes or their products that control commercially important characteristics such as growth or egg production.

FURTHER READING

Bowman, J.C. (1974) *An Introduction to Animal Breeding*. Edward Arnold (Publishers) Limited, London.

Crawford R.D. (ed.) (1990) *Poultry Breeding and Genetics*. Elsevier, Amsterdam.

Hill, W.G., Manson, J.M. and Hewitt, D. (eds) (1985) *Poultry Genetics and Breeding*. Poultry Science Symposium No. 18. British Poultry Science Limited, London.

Leclercq, B. and Whitehead, C.C. (eds) (1988) *Leanness in Domestic Birds*. Butterworths, London.

Mather, C. (1994) Transgenic chicken by DNA microinjection. *Poultry International* 33(6), 16, 18.

Owen, J.B. and Axford, R.F.E. (eds) (1991) *Breeding for Disease Resistance in Farm Animals*. CAB International, Wallingford, UK.

Nutrition and Feeding   8

Food accounts for over 70% of the cost of producing poultry meat and over 60% of the cost of producing an edible egg, so it is important that poultry efficiently digest and use the foodstuffs and nutrients in their diets. The study of nutrition involves an understanding of digestive physiology and biochemistry. However, the application of our knowledge of poultry nutrition must interact with the economic considerations that influence the amounts of nutrients we supply in practical poultry rations.

THE DIGESTIVE TRACT

Digestion of food involves the physical and enzymic breakdown of complex plant and animal material. This breakdown gives chemical units that are small enough to be absorbed though the villi of the gut wall into the blood stream. The digestive tract of poultry (Fig. 8.1) is broadly similar to that of pigs and humans. They are all single stomached (non-ruminant) animals who have a poor ability to digest cellulose and other complex carbohydrates when compared to ruminant animals.

The need for poultry to become airborne, at least on some rare occasions during their domesticated life, has ensured that any evolutionary pressures have kept the weight of their digestive tracts to a minimum compared with their body weights. The digestive tracts of poultry are lighter, shorter and food passes through much more quickly when compared to other non-ruminant species.

Mouth

Poultry have a characteristic horny beak that is usually sharp and pointed although waterfowl have rounded beaks with inward facing ridges to enable food to be obtained underwater. The mouth has an

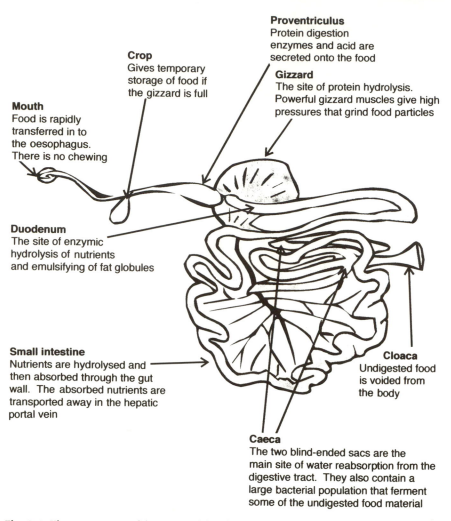

Proventriculus
Protein digestion
enzymes and acid are
secreted onto the food

Crop
Gives temporary
storage of food if
the gizzard is full

Gizzard
The site of protein hydrolysis.
Powerful gizzard muscles give high
pressures that grind food particles

Mouth
Food is rapidly
transferred in to
the oesophagus.
There is no chewing

Duodenum
The site of enzymic
hydrolysis of nutrients
and emulsifying of fat globules

Small intestine
Nutrients are hydrolysed and
then absorbed through the gut
wall. The absorbed nutrients are
transported away in the hepatic
portal vein

Cloaca
Undigested food
is voided from
the body

Caeca
The two blind-ended sacs are the
main site of water reabsorption from the
digestive tract. They also contain a
large bacterial population that ferment
some of the undigested food material

Fig. 8.1. The structure and function of the digestive tract.

inflexible tongue that moves only forwards and backwards and passes
food particles rapidly into the oesophagus. There is no chewing of the
food. There are very few taste buds and, although saliva is secreted,
there is little enzymic breakdown of food in the mouth.

Crop

The crop is an expandable sac that is half way down the length of the
oesophagus. Its main function is food storage. Food particles pass
straight down the oesophagus and bypass the crop when lower parts of

the digestive tract are empty. Surplus food enters and is stored in the crop if the gizzard is full. Short contractions of the crop over the next 6 h then release food back into the oesophagus.

The crop has a resident bacterial population, mostly *Lactobacillus* species, which ferment some carbohydrates in the food for their own purposes. Lactic acid is the by-product of this fermentation so the pH of the feed drops as the food is retained in the crop. A condition called 'sour crop' or 'pendulous crop' occasionally occurs in some birds whose physiological mechanism of food release from the crop becomes blocked. Birds that suffer from sour crop may have very low pH levels in their crop due to the extended length of the bacterial fermentation.

Geese and ducks do not have a true crop but have an enlarged part of the lower oesophagus. This gives waterfowl a similar ability to store food but bacterial fermentation is less compared to other poultry.

Proventriculus

Once the food reaches the end of the oesophagus it enters the proventriculus. The proventriculus is distinguished by its thick glandular mucous membrane. The glands secrete hydrochloric acid and pepsinogen (the precursor of pepsin). The acidity of the food is reduced to a pH that enables pepsin to be formed and to catalyse protein hydrolysis.

Gizzard

The food moves quickly through the proventriculus but is held for a longer time in the gizzard, so most of the pepsin-catalysed protein hydrolysis occurs in this part of the digestive tract. The gizzard is a flattened sphere surrounded by powerful muscles that generate high pressures within. Muscular contractions can physically break up very dense food particles such as whole cereal grains. Particles of grit are retained in the gizzard and they provide an abrasive surface that helps in grinding the food. The submucosa of the gizzard secretes a protein–polysaccharide substance called koilin. The koilin solidifies into short rods when it reaches the acid conditions in the gizzard and the rods cross-link to form a mesh around the gizzard wall. This protects the wall from damage and provides an abrasive surface for the grinding process.

Small Intestine

Once food particles are small enough to be released from the gizzard they enter the duodenum where bile and pancreatic enzymes are secreted. A large range of digestive enzymes that hydrolyse carbohydrates, proteins and fats are also secreted from the brush border of the small intestine. The small intestine of poultry is short and food passes quickly down the tract.

Poultry have developed a means of increasing the time available for nutrient hydrolysis. Waves of muscular contractions, called peristalsis, propel food down the digestive tract. Reverse duodenal peristalsis refluxes some of the contents of the duodenum back against the overall flow of digesta into the gizzard. Reverse peristalsis is particularly important to maximize the hydrolysis and absorption of lipids. The acidity of the gizzard does not provide the correct pH for lipase to act, but the bile salts are not pH dependent and will continue to emulsify fat globules.

Large Intestine

Poultry have a very short colon and rectum but have two large caeca. The caeca are the sites of bacterial fermentation of undigested food material in most non-ruminant livestock. Short chain fatty acids are the products of these fermentations and they are then absorbed into the blood stream. The colon is the major site of water reabsorption in most non-ruminant species. Poultry caeca contain a large bacterial flora but evidence shows that caecal bacterial fermentation makes little or no contribution to the overall nutrient supply of the bird. The main function is water reabsorption that complements the action of the relatively short colon. Reverse peristalsis pushes the undigested feed material that arrives in the colon back to the caeca. The process is selective and moves greater proportions of water than solids. Even some water from the urine that arrives in the cloaca is reclaimed this way and this leaves solid, white urates coating the faeces.

Cloaca

The undigested feed residues accumulate in the cloaca from where they are excreted. Urine is also accumulated in the cloaca and will mix with the faeces before being voided from the bird's body. The digestibility of nutrients within the feed is difficult to estimate in poultry because of this mixing with urine before excretion.

NUTRIENT ALLOWANCES

Detailed recommendations on the requirements of poultry have been made by several expert groups. Most nutrients are required by poultry to maintain their current state (maintenance) and to enable weight gain and egg production (production). Energy and high quality protein are the two most expensive nutrients to supply in a practical poultry ration and hence they are often of greatest concern to poultry nutritionists. The requirements for these two nutrients are dealt with in later sections.

Poultry also have requirements for fatty acids, minerals, vitamins and water. Linoleic acid is the only fatty acid required by poultry: it is needed in very small amounts by growing birds but may be required up to 10 g kg^{-1} of the feed for laying birds to maximize their egg weights.

Minerals are the inorganic part of the diet required by poultry. Calcium, phosphorus, sodium, potassium, magnesium and chloride are required in relatively large amounts although only the first three named may need to be supplemented in practical poultry feeds. These minerals have a variety of functions such as the formation of skeleton and egg shells and the control of osmotic balance in the body, and some act as cofactors of enzymes. Copper, iodine, iron, manganese, selenium and zinc are needed only in trace amounts. There are often, but not always, sufficient amounts naturally present in commonly used foodstuffs.

Vitamins are organic compounds needed in small amounts by poultry. They are usually classified into two groups. Water-soluble

Table 8.1. The metabolizable energy requirements of poultry (MJ day^{-1}).

Need for energy	Primarily affected by	Other influences	Predicted by
Maintenance	Body weight	Activity and body composition of the bird	$0.35\ W^{0.75}$
	Loss of body heat	Effective temperature of the environment, feather cover of the birds and their amount of subcutaneous fat	$(a - bT)\ W^{0.75}$
Production	Weight gain or loss	Proportion of lean meat and fat in the weight change	$(10L) + (56F)$
	Egg mass output	Composition of yolk, albumen and shell in the egg	$(25Y) + (3.6A) + (1.25S)$

a and b are constants that relate to the degree of insulation of the bird. These two constants are difficult to estimate but $a = 0.21$ and $b = 0.0082$ are good values for well-feathered growing or laying chickens. A = *weight of albumen produced (kg); F* = gain of fat (kg); L = weight gain not including the gain of fat (kg); S = weight of shell produced (kg); T = effective ambient temperature (°C); W = body weight (kg); Y = weight of yolk produced (kg).

vitamins have functions of coenzyme transfer in the metabolism of carbohydrates, lipids or amino acids or in catalysing reactions involved in electron transport. Fat-soluble vitamins are absorbed into the body along with dietary fats and are stored in the liver. Vitamin A is involved in the formation of rhodopsin in the eye, the formation of mucus in the skin and in the formation of some steroid hormones. Vitamin D is a precursor of a steroid hormone that governs calcium transport within the body. Vitamins E and K are two other fat-soluble vitamins required by poultry.

Water is also an essential nutrient for all birds. Most practical 'dry' poultry feeds will contain 10–15% water but, in addition, a bird needs to drink about twice the weight of water that it eats of 'dry' food.

Energy

Poultry can derive energy from simple carbohydrates, some complex carbohydrates, protein, oils and fat. Complex carbohydrates such as cellulose cannot be utilized, so a system of describing the available energy content of a poultry feed is needed. Metabolizable energy is the conventional system of describing the available energy content of food and the requirements of poultry (see Table 8.1 and Boxes 8.1, 8.2).

Box 8.1. Energy content of poultry feeds.

Definitions

Gross energy (GE) is the energy liberated when a feed is burnt in oxygen. **Metabolizable energy (ME)** is the GE of a feed less the energy lost in the excreta. ME is the energy that is available to the bird.

e.g. Wheat: GE = 16.9 MJ kg^{-1} ME = 13.1 MJ kg^{-1}

Measurement of Available Energy

Apparent metabolizable energy (AME)

AME (MJ kg^{-1} of food) = GE intake kg^{-1} – GE excreted kg^{-1}

AME is usually measured using group-housed birds given a nutritionally adequate diet for 7–8 days.

True metabolizable energy (TME)

TME (MJ kg^{-1} of food) = GE intake kg^{-1} – GE excreted kg^{-1} + endogenous energy loss (kg)

A small amount (30–50 g) of a test food is given to previously starved adult birds. Unfed birds give data on endogenous energy loss. Nutritionally imbalanced feeds can be tested.

Prediction Equations

The chemical composition of a mixed food can be used to predict its ME. A large number of prediction equations are available. The one given below is used for the prediction of the ME of feeds that are marketed within the EU.

ME (MJ kg^{-1}) = 0.3431 (% fat) + 0.1551 (% crude protein) + 0.1301 (% total sugar)
                + 0.1669 (% starch)

Box 8.2. Prediction of energy requirements of poultry: a worked example.

Bird description: A flock of laying domestic fowl have a mean body weight of 2.05 kg and a mean egg mass output of 59.3 g per bird per day. The egg composition is 32% yolk, 56% albumen and 12% shell. The birds are gaining weight at a rate of 0.7 g day^{-1}. Fat accounts for 40% of this weight gain. The birds are well feathered and kept in groups at an ambient temperature of 18°C (see Box 8.1 for prediction equations).

Purpose of energy	Factors that affect the requirement	Calculations	Amount (MJ day^{-1})
Maintenance	Body weight (2.05 kg)	$0.35 \times (2.05^{0.75}) = 0.35 \times 1.713$	0.60
	Heat loss (at 18°C)	$(0.21 - (0.0082 \times 18)) \times 2.05^{0.75}$	0.106
		Requirement for maintenance =	0.706
Production (weight gain) (0.0007 kg day^{-1})	40% fat gain (0.0007×0.4)	$(0.0007 \times 0.4) \times 56$	0.016
	60% non-fat gain (0.0007×0.6)	$(0.0007 \times 0.6) \times 10$	0.004
Production (egg production) (0.0593 kg day^{-1})	32% yolk output (0.0593×0.32)	$(0.0593 \times 0.32) \times 25$	0.474
	56% albumen (0.0593×0.56)	$(0.0593 \times 0.56) \times 3.6$	0.119
	12% shell output (0.0593×0.12)	$(0.0593 \times 0.12) \times 1.25$	0.009
		Requirement for production =	0.623
		Total energy requirement =	1.329

Energy is required in varying amounts for all metabolic purposes, so a deficiency of energy affects most aspects of productive performance of poultry. If the available energy concentration of the diet (the number of MJ per kilogram of feed) is changed then poultry maintain a constant energy intake by changing their feed intakes. The changes in feed intakes are not perfect and high energy concentrations in a feed usually result in slightly higher energy intakes. However, the energy concentrations used in practical poultry feeds are not critical and can be allowed to vary. The available energy concentration that gives the lowest cost for a unit of energy is often used.

The productive performance of commercial poultry is greatest if free access to feed is given, whatever its energy concentration. Broiler chicken breeding stock are an exception. Restriction of energy intakes during rearing, up to 50% of ad libitum intakes, improves egg numbers particularly in the early part of the laying period. Restriction of energy intakes during lay improves fertility and increases the hatchability of the fertile eggs.

Protein

Feed proteins are a mixture of approximately 22 amino acids. Eleven are essential amino acids. Essential amino acids cannot be synthesized at all or are synthesized too slowly by the body and need to be supplied in the diet. The remainder of the 22 are non-essential and can be produced from other amino acids.

A correct balance of essential amino acids is required in the diet to meet the requirements of the bird. Additionally, a correct balance of essential to non-essential amino acids is also necessary. Diets are poorly utilized if they have minimum levels of essential amino acids but have non-essential amino acids in excess. There is therefore an ideal balance of dietary amino acids for poultry. The ideal balance differs according to the productive output of the bird. High levels of lean meat deposition require relatively high levels of lysine. High levels of egg output or feather growth require relatively high levels of sulphur amino acids (Table 8.2 and Boxes 8.3, 8.4).

Calculation of amino acid requirements

The requirement for amino acids depends on the individual birds' need for maintenance and its need for live weight gain or egg output. The

Table 8.2. The ideal balance of amino acids for poultry (relative to lysine = 1.00).

| Amino acid | Domestic fowl | | Growing turkeys | Growing ducks |
	Growing flocks	Egg laying flocks		
Lysine	1.00	1.00	1.00	1.00
Arginine	1.05	1.06	1.10	1.00
Glycine and serine	1.31	0.78	1.27	1.27
Histidine	0.40	0.25	0.38	0.43
Isoleucine	0.72	0.78	0.69	0.77
Leucine	1.25	1.14	1.16	1.30
Methionine and cystine	0.75	0.86	0.75	0.75
Phenylalanine and tyrosine	1.21	1.25	1.08	1.20
Threonine	0.63	0.69	0.68	0.66
Tryptophan	0.18	0.24	0.17	0.19
Valine	0.79	0.87	0.76	0.89

> **Box 8.3.** Calculation of the ideal level of amino acids in poultry diets: worked example 1.
>
> Flock description: A flock of 10-week-old turkeys have a mean weight of 3.78 kg. Their current mean growth rate is 86 g day^{-1} with a standard deviation of 9.2 g day^{-1}. An initial study of the possible diet indicates that **lysine** is the first limiting amino acid and that the most profitable level of supply is to provide for 90% of the flock's requirement. The turkeys are eating 170 g of food per day.
>
> Calculation for lysine requirement (using the coefficients from Table 8.3):
>
> Maintenance requirement (mg day^{-1})   = 19×3.78 =   71.8 mg
> *(3.78 kg body weight)*
>
> Production requirement (mg day^{-1})   = 21.16×98.0 =   2073.7 mg
> *(A gain of 1.3 standard deviations above the mean is required to meet 90% of the individual's growth potentials. This is calculated as $86 + (1.3 \times 9.2) = 98.0$ g day^{-1})*
>   TOTAL   =   2145.5 mg
>
> Level in feed:
> The turkeys need 2145.5 mg in 170 g of feed   =   2145.5/170   =   12.6 g kg^{-1} of feed
>
> Calculations for other amino acids:
>
> Methionine and cystine:
> Table 8.2 indicates the ideal ratio of methionine plus cystine to lysine is 0.75.
>
> Level of methionine and cystine in feed   = 12.6×0.75 =   9.4 g kg^{-1} of feed
>
> Arginine:
> Table 8.2 indicates the ideal ratio of arginine to lysine is 1.10
>
> Level of arginine in feed   = 12.6×1.10 =   13.9 g kg^{-1} of feed
>
> This procedure is used to calculate the levels of all other essential amino acids.

requirement of an individual can be estimated by calculation using the coefficients given in Table 8.3.

If a diet is deficient in one amino acid then additions of this amino acid will result in an increased productive output in the birds (Fig. 8.2). Productive output is the total weight of eggs in laying birds or live weight gain in growing birds. The increase in output will continue until the maximum genetic potential of the bird is reached or the amino acid is no longer first limiting.

The genetic potential for productive performance varies between individual birds in a flock. Therefore, a single diet can never supply the correct level of an amino acid for each bird within a flock. The increase in egg output or live weight gain of a flock of birds with increasing amino acid intakes is therefore a curved response. The curve only slowly reaches a point where there is no further increase in productive output.

Box 8.4. Calculation of the ideal level of amino acids in poultry diets: worked example 2.

Flock description: A flock of laying hens have a mean live weight of 2.05 kg and there is no overall weight gain or loss. Their mean egg output is 58.7 g day^{-1} with a standard deviation of egg output of 4.18 g day^{-1}. An initial study of the possible diet indicates that **methionine plus cystine** are first limiting and the economic level of addition is to meet 95% of the flock's requirement. The birds are eating 114 g day^{-1}.

Calculation for methionine and cystine requirement (using the coefficients from Table 8.3):

Maintenance requirement (mg day^{-1}) (2.05 kg body weight)	=	13.0×2.05	=	26.7 mg

Production requirement (mg day^{-1})   =   10.15×65.4   =   663.8 mg
(*An egg output of 1.6 standard deviations above the mean is required to meet 95% of the individual's potentials. This is calculated as 58.7 + (1.6×4.18) = 65.4 g day^{-1}*)

                                                                                TOTAL   =   690.5 mg

Level in feed:
The laying hens need 690.5 mg in 114 g of feed   =   690.5/114   =   6.06 g kg^{-1} of feed

Calculations for other amino acids:

Lysine:
Table 8.2 indicates the ideal ratio of methionine plus cystine to lysine is 0.86.

Level of lysine in feed   =   6.06×(1.0/0.86)=   7.05 g kg^{-1} of feed

Arginine:
Table 8.2 indicates the ideal ratio of arginine to lysine is 1.06

Level of arginine in feed   =   7.05×1.06   =   7.47 g kg^{-1} of feed

This same procedure is used to calculate the levels of all other essential amino acids.

Figure 8.3 shows the response of a flock to an increase of methionine plus cystine when these sulphur amino acids are first limiting in the diet. The three points show the levels of amino acid that are necessary to get 50, 90 or 95% of the maximum productive output possible from the flock. Only small increases in the intake of methionine and cystine are needed to get relatively large increases in egg output between 50 and

Table 8.3. Coefficients for calculating the requirement of two limiting amino acids.

Amino acid	Laying domestic fowl		Broiler chickens		Turkeys	
	a	b	a	b	a	b
Lysine	9.99	73	14.86	82	21.16	19
Methionine plus cystine	10.15	13	11.59	41	11.28	98

Amino acid required (mg day^{-1}) = a×productive output (g day^{-1}) + b×live weight (kg)

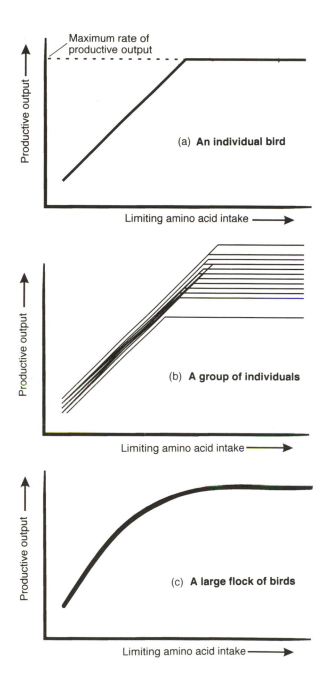

Fig. 8.2. A model of the response of poultry to increasing intakes of a limiting amino acid.

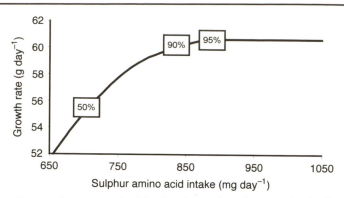

Fig. 8.3. The predicted growth of flocks of meat-line domestic fowl when given diets that vary in their methionine plus cystine (sulphur amino acid) content.

The three marks on the curve show the sulphur amino acid intake where 50, 90 and 95% of the birds in the flock are at or above their requirement. Only birds that are below their required intakes will increase their growth rate when given more sulphur amino acids and when these are first limiting amino acid in the diet. Notice that above the 90% mark, a large increase in sulphur amino acid intake is needed to get a relatively small increase in the mean growth rate of the flock.

90% of the maximum productive output. After that, there are continuously diminishing responses to increases in sulphur amino acid intakes above this point.

The extra value of the egg output will most probably outweigh the cost of increasing the intake of a limiting amino acid at low dietary levels. The economic merit of increasing the limiting amino acid intake at higher dietary levels is less certain. This depends on the cost of adding an extra unit of the amino acid in the diet compared to the value of the extra product obtained. As a rule, the most profitable point of supplementation of the usual limiting amino acids in poultry feeds is to meet between 90 and 95% of the flock's requirement. These points can

Table 8.4. Using linear-programming feed formulation computer programs.

Information needed to be put in to the program	Information given by a feasible solution of the computer program
1. Nutrient requirements of the class of poultry 2. List of the available feedstuffs and their price per tonne. 3. The nutrient compositions of each of the feedstuffs. 4. The minimum, maximum or exact levels of a specific feedstuff that is required in the final diet.	1. The levels of each feedstuff used in the optimal solution. This is the formulation that meets all the constraints at the lowest cost. 2. The price of the optimal solution. 3. The price that any unused feedstuff would need to be for it to be used in a best cost ration. 4. The reduction in the cost of the ration if any of the nutrient specifications were reduced by one unit.

be estimated by measuring the variability of the output of the flock and expressing it as a standard deviation. Multiplying the standard deviation by 1.3 or 1.6 and adding this value to the mean output of the flock gives the 90 and 95% points respectively.

Prediction of Food Intake

Poultry eat a daily amount of food that is equivalent to approximately 5% of their body weight. They eat the amount of food that approximately meets their energy requirements. Prediction equations of food intakes are usually based on an estimate of the requirement of poultry for metabolizable energy (see earlier section).

The food intakes of poultry may be further altered by factors that are not directly related to the nutrient content of the feed. Intakes of pelleted feeds are typically 5–8% greater than when the birds are given the same feed in meal form. Short day lengths result in low food intakes, especially in young birds. Feed troughs with poor access or high stocking

Table 8.5. Typical nutrient specifications used for poultry feeds.

a) Domestic fowl

Nutrient	Units	Growing meat-line birds			Meat-line breeding stock			Egg laying strains		
		0–18 days	19–35 days	35 days on	0–5 weeks	6–22 weeks	Laying	0–5 weeks	6–18 weeks	Laying
Metabolizable energy	MJ kg⁻¹	12.6–13.0	12.8–13.1	13.0–13.4	11.5–12.0	11.0–11.6	11.5–11.8	11.5–12.5	11.3–12.0	10.2–11.7
Crude protein	g kg⁻¹	210–240	200–230	180–220	170–200	150–180	140–170	180–220	140–180	150–190
Lysine	g kg⁻¹	> 13	> 12	> 11	> 9	> 7	> 6	> 9	> 7	> 7
Methionine plus cystine	g kg⁻¹	> 9	> 8	> 7	> 7	> 5	> 5	> 6	> 5	> 6
Linoleic acid	g kg⁻¹	–	–	–	> 12	> 10	> 12	> 10	> 8	> 12
Calcium	g kg⁻¹	9–14	8–13	8–13	9–14	12–18	28–35	9–14	8–13	36–42
Phosphorus	g kg⁻¹	6–12	5–11	5–11	6–12	5–8	5–8	7–12	6–11	5–12
Sodium	g kg⁻¹	1.6–2.5	1.6–2.5	1.6–2.5	1.6–2.5	1.6–2.5	1.6–2.5	1.6–2.5	1.6–2.5	1.6–2.5

b) Other classes of poultry

Nutrient	Units	Turkeys		Ducks		Geese		Quail	
		0–3 weeks	12 weeks on	0–3 weeks	6 weeks on	0–3 weeks	7 weeks on	0–3 weeks	4 weeks on
Metabolizable energy	MJ kg⁻¹	11.8–12.2	12.4–13.0	11.7–12.2	12.1–12.5	11.0–12.0	11.3–12.2	12.0–13.2	12.0–13.0
Crude protein	g kg⁻¹	260–300	160–210	210–240	150–170	160–190	100–140	230–270	180–220
Lysine	g kg⁻¹	> 16	> 11	> 12	> 8	> 9	> 5	> 14	> 13
Methionine and cystine	g kg⁻¹	> 10	> 7	> 8	> 6	> 8	> 5	> 9	> 8
Linoleic acid	g kg⁻¹	> 14	> 9	> 12	> 8	–	–	–	–
Calcium	g kg⁻¹	10–14	9–14	8–11	7–12	6–10	6–10	9–13	8–13
Phosphorus	g kg⁻¹	8–13	7–13	6–10	5–11	5–9	5–9	7–12	7–12
Sodium	g kg⁻¹	1.6–2.5	1.6–2.5	1.6–2.5	1.6–2.5	1.6–2.5	1.6–2.5	1.6–2.5	1.6–2.5

densities reduce feed intakes. Providing food as a wet mash gives a short-term stimulus to food intakes. High moisture foods are prone to bacterial spoilage and so they are not generally used in large poultry enterprises.

PRACTICAL FEED FORMULATION

Feed Formulation Using Linear Programming Techniques

A mixture of four or five different foodstuffs could probably meet all the nutrient requirements of most classes of poultry. The formulation with the best, not necessarily the lowest, cost is more likely to include up to 12 different foodstuffs.

The objective of the formulation of proprietary poultry feeds is to provide correct nutrient intakes at the best possible cost. Linear programming is a matrix algebra technique designed to find the solution to this problem. A series of calculations, called iterations, are used to produce feasible solutions to the problem. Over 25 separate iterations may be needed to solve a typical poultry feed formulation. Computers are essential to allow linear-programming to be used regularly (see Table 8.4).

Composition of Poultry Feeds

Many combinations of foodstuffs could be permutated that meet the constraints of a typical poultry feed specification (Tables 8.5 and 8.6). In practice, most poultry feeds have a very similar composition. Formulations for commercial poultry diets tend to be low in fibre and are cereal

Table 8.6. Types of feedstuffs used in practical poultry feeds.

Feedstuff classification	Examples	Feedstuff classification	Examples
Cereals	Wheat	Protein concentrates	Soyabean meal
	Maize		Groundnut meal
	Barley		Fish meal
	Rice		Meat and bone meal
Cereal by-products	Wheatfeed	Oils and fats	Soyabean oil
	Rice bran		Rapeseed oil
	Maize gluten meal		Fish oil, tallow
	Maize germ meal	Minerals and vitamins	Limestone
			Dicalcium phosphate
			Salt

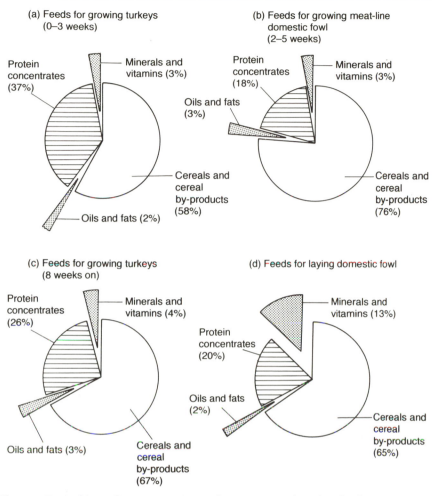

Fig. 8.4. Typical ingredient composition of some practical poultry feeds.

based. Cereals often comprise between 50 and 75% of a poultry diet (Fig. 8.4). They supply a high proportion of starch that is frequently the lowest cost form of available dietary energy. Animal fats or vegetable oils may also be used as a source of dietary energy up to a maximum inclusion of 6% (see Fig. 8.5). Above this level it often becomes difficult to form hard enough pellets or to mechanically move the sticky feed when it is not pelleted.

Cereals may also contribute up to 50% of the crude protein required in the feed, but this protein is usually deficient in essential amino acids. Lysine is particularly deficient in the protein of cereals. Concentrated sources of protein must therefore be used to increase the amino acid content of the feed. Oilseed meals are generally by-products of the oil-

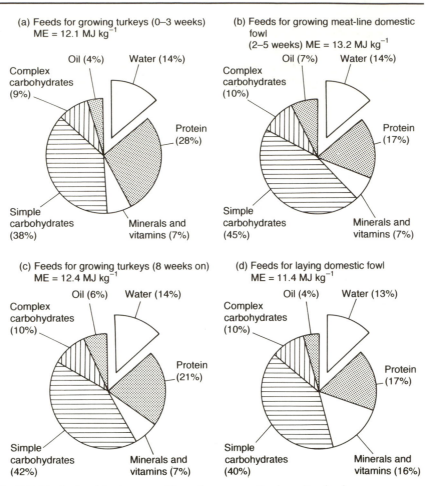

Fig. 8.5. Typical nutrient composition of some practical poultry feeds.

crushing industry. They are often low cost, protein rich feeds. Examples are soyabean meal, rapeseed meal, sunflower seed meal, groundnut meal and sesame seed meal. Soyabean meal is the most commonly used oilseed meal in the world. This protein concentrate contains almost 50% crude protein and has a high lysine content. By-products of the animal production industry, for example meat and bone meal, fish meal and poultry offal meal, are also used as protein concentrates.

Some minerals such as calcium, phosphorus and sodium are needed in relatively large amounts by poultry. Poultry have unusually large requirements for calcium during periods of egg production. Minerals and vitamins are provided, usually only in part, by the cereals and protein concentrates in the ration. Poultry diets are often supplemented with other sources of minerals. Limestone is a source of available calcium and dicalcium phosphate provides both calcium and phosphorus. Common salt provides sodium. Proprietary vitamin and trace element mixes are usually included in poultry feed formulations. Some pure or highly concentrated forms of individual amino acids may also be added.

High dry matter, high nutrient concentration feeds have many advantages that make them suitable to be used in large poultry enterprises. Storage and movement of these feeds are easier and their distribution around the poultry house is much simpler.

FURTHER READING

Boorman, K.N. and Freeman, B.M. (eds) (1975) *Digestion in the Fowl* Proceedings of the 11th Poultry Science Symposium, British Poultry Science Ltd.

Cole, D.J.A. and Haresign, W. (eds) (1989) *Recent Developments in Poultry Nutrition*. Butterworths, London.

Fisher, C. and Boorman, K.N. (eds) (1986) *Nutrient Requirements of Poultry and Nutritional Research*. Butterworths, London.

Larbier, M. and Leclercq, B. (1992) *Nutrition and Feeding of Poultry*. Nottingham University Press, Loughborough, UK.

Leeson, S. and Summars, J.D. (1991) *Commercial Poultry Nutrition*. University Books, University of Guelph, Canada.

Lonsdale, C. (1989) *Straights. Raw Materials for Animal Feed Compounders and Farmers*. Chalcombe Publications, Marlow, UK.

NRC (1994) *Nutrient Requirements of Poultry*, 9th revised edn. National Research Council, Washington.

Scott, M.L., Nesheim, M.C. and Young, R.J. (1982). *Nutrition of the Chicken*, 3rd edition. M.L. Scott and Associates, Ithaca, New York.

Wiseman, J. and Cole, D.J.A. (eds) (1990) *Feedstuff Evaluation*. Butterworths, London.

Housing and Welfare   9

HOUSING AND ENVIRONMENTAL CONTROL

Many poultry flocks are kept in houses in which important environmental variables, particularly temperature and light, are controlled. Controlled environment houses can give accurate control over light, and can also control temperature if outside temperatures are above or below those required inside the house. These types of houses are frequently used in temperate climates where they can significantly improve the productive performance of poultry. Poultry buildings in hotter climates are often much lower cost structures with open sides, that give less accurate control of temperature and light.

A disadvantage of stocking poultry intensively within buildings is that high concentrations of environmental contaminants can accumulate within the house. The purpose of this section is to explain the physiological effects on poultry of changes in the main environmental variables and to describe briefly how they are controlled in practical poultry systems.

Temperature

The response of birds

Poultry are homeotherms that attempt to maintain deep body temperatures around 41°C. The body temperature of poultry is usually greater than the ambient temperature, so heat will be continually lost to the environment. Heat is lost by a mixture of four different mechanisms; these are convection, conduction, radiation and water evaporation.

Convection, conduction and radiation are often called sensible heat losses. Sensible heat losses depend mostly upon the size of the bird, the temperature of its environment and the quality of its own feather cover: values range from 0.7 watts per bird for a day-old chick to 8 watts per bird for a well feathered laying hen, and up to 20 watts per bird for a 6 kg

106°F

turkey at normal temperatures. Sensible heat output rises when house temperatures decrease and birds with poor feather cover have a greater increase in sensible heat loss per degree drop in temperature.

Poultry can alter their sensible heat losses to control their body temperatures. For example, a bird that is too hot will divert blood flow to the comb and wattles on the head and will also increase blood flow to the legs. Birds will make postural changes such as resting with wings raised and legs lying away from the body to increase their convection heat losses. Birds within a flock will space out to attempt to increase the air flow around themselves and to reduce conduction and radiation heat gains from other birds. The birds will also avoid resting in direct sunlight which would increase their heat gain from radiation. Conversely, birds that are too cold will reduce blood flow to body extremities. They will huddle together, avoid air draughts and gather round any sort of radiative heat source such as direct sunlight or a brooder lamp.

Evaporative heat loss is called insensible heat loss. Water evaporation occurs continuously as birds breathe but this type of heat loss is small at lower temperatures; it accounts for about 0.6% of a bird's body heat output. Birds start to pant when their body temperatures rise above 41°C. They increase their respiration rates by shallow panting so increasing their evaporative heat losses. Birds kept at 35°C can lose about 2% of their total heat output by water evaporation from their respiratory tracts. The effectiveness of evaporative heat loss depends not only on the air temperature but also on the relative humidity of the air. Panting requires extra energy and so some of the birds' energy intake needs to be used for this purpose. Digestion of feed requires energy expenditure and waste heat release. The bird reduces its voluntary feed intake to reduce this form of body heat production. The energy intake that remains available for weight gain or egg output is therefore reduced and the birds will not reach their genetic potential in these parameters.

Birds kept at low ambient temperatures need to metabolize stored body energy to maintain their deep body temperatures close to 41°C. The energy requirements of the birds are increased and they will increase their voluntary feed intakes to meet these requirements. Growth rates or egg outputs are unlikely to be changed at these lower temperatures and so the efficiency of feed utilization will be decreased (Fig. 9.1). Growth rates or egg outputs only decrease when the temperature is so low that the birds cannot physically increase their voluntary feed intakes to meet their increased energy requirements fully.

The best temperature to keep fully feathered poultry is difficult to estimate. Temperatures between 18°C and 24°C are generally preferred, but this depends on the relative prices of feed, poultry meat and the cost of providing housing and heating.

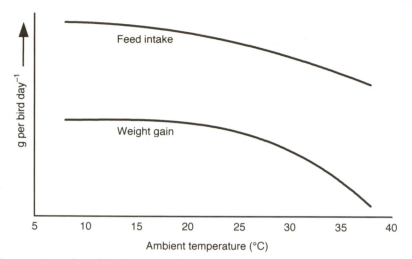

Fig. 9.1. Growth and feed intakes of meat-line domestic fowl kept at different temperatures.

Young chicks have few feathers and little subcutaneous fat. They therefore have little body insulation and they have a proportionally high surface area compared to adult birds. They are also less able to metabolize body fat stores quickly to produce heat. Day-old birds are therefore more susceptible to low temperatures compared to older birds. An air temperature of 32°C is correct for newly hatched chicks. The correct rearing temperature decreases gradually as the birds grow and get a complete feather cover.

Temperature control in practical systems

In temperate conditions and at usual stocking densities, a flock of birds kept within an insulated poultry house gives off more waste heat than is lost through the building structure. The temperature within the house could quickly rise to a level that would be lethal to the birds unless cooler air is brought in from outside the building. A major purpose of a ventilation system is thus to regulate temperatures inside a poultry house (Fig. 9.2).

A robust temperature sensor is an important part of a practical ventilation system. Two types of sensors are used. Capillary thermostats have fluid in a bulb and changes in vapour pressure are detected. Electronic thermostats measure the changes in the resistance of a wire due to temperature changes. A number of sensors can be linked together and placed throughout a large house to give an average temperature reading.

The information from the temperature sensors can be used to control electrical fans in a powered fan system. A method for calculating the number of powered fans needed in a controlled environment poultry house is given in Box 9.1. Powered fan systems may either push cool air into the house (a positive pressure system) or extract warm air from the house (a negative pressure system). Air inlets need to be placed away from the air outlets to ensure that there is good mixing of incoming cool air with the existing warm air. The inlets and outlets may be on opposite sides of the house or one of these types of vents may be in the ridge of the house.

Simple control systems turn fans off or on once the house temperature is below or above a set temperature. Starting or stopping groups of fans at slightly different temperatures, or gradually changing the speed of variable speed fans, gives more accurate temperature control.

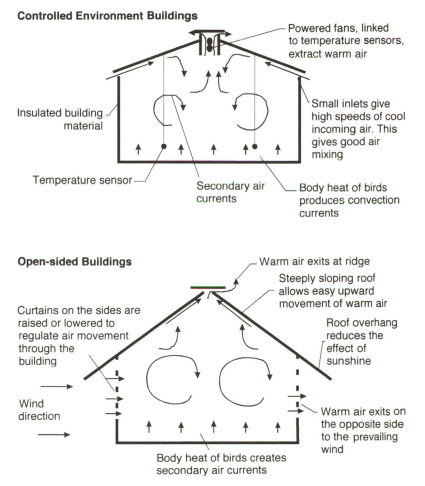

Fig. 9.2. Temperature control in poultry houses.

Box 9.1. Calculation of the number of powered fans needed in a controlled environment poultry house.

The number of fans should be able to provide the maximum ventilation rate needed which can be calculated by the following approximate equation:

Maximum ventilation (m^3 h^{-1}) = $S/(T \times 0.35)$

S = sensible heat output of the flock (watts)
T = Maximum temperature difference required between the inside and outside air.

Example: A well insulated poultry house will contain 20,000 laying hens with a mean body weight of 2.1 kg. The birds will be well feathered and their estimated maximum sensible heat output will be 8 watts per bird early in the laying period but it is expected that feather cover will deteriorate towards the end of the lay and so their sensible heat output will rise to 12 watts per bird. The house will be fitted with 630 mm diameter powered fans that each have an approximate output of 10,000 m^{3}h^{-1}. A maximum temperature difference of 4°C between inside and outside air is required.

Maximum ventilation (m^3 h^{-1}) = (20,000 $\times$ 12) / (4 $\times$ 0.35) = 171,429 m^3 h^{-1}

Number of 630 mm fans to install = 171,429 / 10,000 = 17.14

(Round up to 18)

A total of 18 fans should be installed in the house to provide for the maximum ventilation rate.

The amount of wind that hits the side of open-sided buildings has a major influence on the amount of outside air movement into the house, so the siting of the house is important in maximizing the effect of prevailing winds. Some open-sided houses may additionally use some powered fans to help the movement of air into the house. Temperature can also be controlled approximately in open-sided houses by raising or lowering curtains or shutters at the sides of the house to regulate air movement into the house.

If outside air temperatures are very high then some cooling can be achieved by water evaporation. Incoming air can either be passed through wetted pads or through a spray of water. Efficient evaporative cooling systems can reduce incoming air temperatures by as much as 3°C. The process involves increasing the relative humidity of the incoming air, which unfortunately decreases the bird's own ability to reduce its body temperature by its own evaporative heat loss.

When day-old poultry flocks are first placed in a poultry house they have a low waste heat output compared to the volume of air within the house, so additional heat is usually needed. The air in the house can all be heated to 32°C or heaters can be used to give localized areas where 32°C is provided, with cooler temperatures tolerated elsewhere in the house. The heaters used in this latter type of system must have a high radiant heat output for them to be cost-effective. Supplementary heat may be required in a poultry house until the birds have a good feather cover.

Light

Light greatly affects the behaviour and productive performance of poultry so its control is a second major advantage of the use of controlled environment poultry houses. Natural light intensity and day length variations are not always suitable for year-round poultry production systems. Controlled environment buildings can successfully stop natural daylight from entering the house. Open-sided buildings will use supplementary artificial light to supplement the inevitable influence of natural daylight to maximize the productive efficiency of the flock.

Response of the bird
Light intensity

Variations in light intensity (brightness) are detected by the eyes of the bird. The eyes of poultry are generally similar to those of mammals and they consist of a lens that directs light on to a light sensitive surface, the retina. The lens and the retina are separated by a watery vitreous humour. The retina surface is populated with photoreceptors shaped as rods or cones. Rods contain rhodopsin and are primarily of use in low light intensity. Cone shaped photoreceptors are primarily used in bright light conditions. Domesticated poultry have relatively fewer rods than man. This explains their general lack of activity in dark or low light intensities. The reluctance to move in dim light can be used to advantage in the management of birds. Catching birds is much easier if the lights are dimmed because the birds are much less excitable and tend not to move away from the catchers.

Dim light intensities decrease the activity, and hence the energy expenditure, of birds. Practical broiler chicken houses therefore often use low light intensities so that growth rates and feed utilization efficiencies are maximized. Light intensity should not be reduced too far; otherwise the stockworkers cannot see within the house. A bright light intensity should be used for day-old stock so that the birds are more active and find their way to feeders and drinkers within the house. A minimum light intensity of 10 lux is recommended for laying birds.

Day length

Birds mostly rest during periods of darkness and are more active during periods of light. Activity, especially feeding and drinking, is greatest shortly after lights come on and shortly before lights go off. Changes in day length have major effects on the photoperiodic response of egg laying birds. The egg laying responses of laying birds to different lighting programmes are described in detail in Chapter 4.

Continuous light gives the maximum feeding time so would generally give young growing birds the maximum opportunity for growth. This is particularly important for newly hatched chicks where high feed intakes increase the survival rates of a flock and improve their growth rates. However, birds reared on continuous light may panic if there is an unavoidable period of dark, for example during a power cut. It is therefore preferable to have a lighting programme that includes at least a short period of darkness each 24 h.

Light source

Poultry detect a similar range of wavelengths to humans but their eyes are more sensitive to red and green wavelengths. If equal light intensities of each light wavelength are given then blue wavelengths are less effective in providing light to the birds. Experiments have shown that domesticated poultry may be less active and less aggressive in blue light but this is probably just a response to a lowered effective light intensity. Single wavelengths are obtained in practice by filtering out all the other wavelengths from a white light. A lowered intensity of white light usually reduces aggression by the same amount as a similar light intensity of a single wavelength.

Poultry can detect high rates of flickering in light sources. Humans can only detect flickering rates of less than about 70 s^{-1} and faster rates are only perceived as continuous light. Poultry can detect flickering rates of 140 s^{-1} and light from fluorescent lamps will be seen as flickering. The aggressiveness of some poultry flocks can be reduced by changing from tungsten lights to fluorescent lights. The flickering light source may be perceived by the birds as a general lower intensity.

Light control in practical systems

Most controlled environment houses do not rely on using any natural daylight. Light is relatively inexpensive to provide artificially and translucent panels or windows are generally expensive to install and can markedly increase the heat loss from the building.

Light is obtained from conventional GLS (General Light Source) lamps, fluorescent tubes or compact fluorescent lamps. Fluorescent light sources are, in comparison to GLS lamps, expensive to install and expensive to dim but thereafter have low running costs. A relatively even light intensity is needed throughout a poultry house, so a large number of low wattage lamps need to be placed throughout the house. High wattage bulbs may give too high a light intensity at some spots in the house and this may predispose the birds in this area to cannibalism.

Flocks of growing poultry need a bright light intensity up to a few weeks of age followed by low light intensity later in the growing period. The light intensity in these houses therefore needs to be changed during the growth period. Dimming is achieved by linking the electrical lighting circuits either to a variable voltage transformer or to an electronic dimming device. The lighting circuits in poultry houses need to be switched on and off at least once each day. A time switching device is therefore included in the electrical lighting circuits. These devices are either electronic or a mechanical operation.

Accurate control of day length is essential when juvenile birds are being reared for later egg production. Light seepage into the house can interfere with the artificial lighting programmes being given. Both air inlets and outlets therefore need to be protected by light baffles to stop outside daylight entering the house.

Environmental Pollutants

Only a little outside air needs to be brought into a poultry house to control the inside temperature if outside temperatures are cold. Cold weather can therefore result in the buildup of gases, dust particles and water vapour within the house. These air pollutants can have detrimental effects on the birds' health and productive performance. Carbon dioxide, ammonia, methane and hydrogen sulphide are the main gases produced by bird respiration or decomposition of their excreta. Carbon monoxide can accumulate if a gas burner is faulty or is insufficiently supplied with oxygen. Water vapour is a waste product of bird respiration and it also evaporates from excreta and the spillage from drinkers. Dust particles are mostly derived from skin and feather debris but some feed and bedding material particles also occur.

Response of birds

Ammonia is the gas that is most likely to be toxic to poultry in practical systems. It is produced when uric acid in the birds' excreta decomposes. The moisture content of the excreta determines the amount of ammonia produced. Ammonia is an irritant to birds. Levels as low as 5 ppm are detectable by humans and have an unpleasant smell. Levels above 50 ppm cause eye inflammations in poultry and reduce the feed intakes and growth of birds. High ammonia concentrations also increase the risk of respiratory disease in poultry flocks.

Hydrogen sulphide has a characteristic smell of rotten eggs and is produced by the anaerobic decomposition of excreta. Hydrogen sulphide is an irritant and asphyxiant and levels greater than 10 ppm are thought to be dangerous to poultry. Concentrations within a poultry house can soar when stored poultry muck is agitated within the house.

Carbon dioxide and methane are less toxic to poultry. Normal air contains about 300 ppm of carbon dioxide but concentrations can rise to 10–20 times that in controlled environment houses. It may cause panting at very high concentrations. Methane is produced from anaerobic decomposition of excreta. It is lighter than air, so it accumulates away from the birds next to the ceiling of the house. High concentrations increase the explosion risk in the poultry house.

High water vapour levels can increase the water content of the floor litter, which then increases the incidence of leg deformities and skin blemishes in the flock. High levels of humidity coupled with low insulation levels in the building structure can also lead to condensation and deterioration of the building fabric.

Dust concentrations in the air may reach up to 10 mg m^{-3} in some poultry houses. The dust is a high protein material because it is mostly derived from particles of body feathers and skin. The effect of high dust levels on poultry is poorly understood but it is probably highly correlated to the incidence of respiratory disease in poultry flocks. High dust levels damage electrical and electronic equipment within poultry houses and are highly allergenic to some stockworkers.

Air pollutant control in practical systems (Fig. 9.3)

Environmental pollutants are being continuously produced within a poultry house and the rate of production is approximately constant. A practical ventilation system must ensure that there is a continuous, or at least frequent, supply of incoming air. This is called a minimum ventilation rate. A practical rule is that the ventilation system should bring a volume of air into the house each 30 minutes that equals the air volume of the whole house. Sensors that measure ammonia and the relative amount of water vapour in the air can be placed inside a poultry house. Ventilation will be increased above the set minimum rate when there are high levels of these environmental pollutants.

WELFARE

Poultry have obtained some benefits from domestication; they are protected by humans from predation and they may be provided with shelter and feed. However, in many production systems, the process of domestication entails confinement and the restriction of natural behaviour.

There is widespread concern over the well-being of farm animals and a wish to protect animals from abuse and neglect. Abuse is the deliberate causing of suffering. Neglect is the less systematic causing of

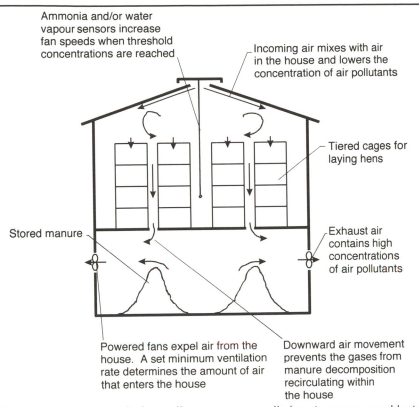

Ammonia and/or water vapour sensors increase fan speeds when threshold concentrations are reached

Incoming air mixes with air in the house and lowers the concentration of air pollutants

Tiered cages for laying hens

Stored manure

Exhaust air contains high concentrations of air pollutants

Powered fans expel air from the house. A set minimum ventilation rate determines the amount of air that enters the house

Downward air movement prevents the gases from manure decomposition recirculating within the house

Fig. 9.3. Practical control of air pollutants in a controlled environment caged laying house.

suffering due to human idleness or ignorance. Good stockworkers generally pride themselves that they completely avoid either type of suffering to poultry. Many countries have laws or welfare codes that protect poultry from suffering.

There is also criticism of normal methods of intensive livestock production, including the methods of poultry production. The Farm Animal Welfare Council consider that there are five basic freedoms that should be given to farm animals (see Box 9.2). Some well-managed poultry production systems provide most, but often not all, of these freedoms. The following sections deal with these five freedoms and give a few examples of poultry production methods where there is welfare criticism.

Freedom from Hunger and Thirst

Most poultry production systems allow the birds free access to feed and water. The main welfare considerations are whether there is adequate

---

Box 9.2. The five freedoms for farm animals (Farm Animal Welfare Council 1992).

1. Freedom from hunger and thirst.
2. Freedom from discomfort.
3. Freedom from pain, injury or disease.
4. Freedom to express normal behaviour.
5. Freedom from fear and distress.

---

space available for all birds to feed and drink without excessive crowding or blocking of access to weaker birds. Most codes of welfare state minimum allowances of feeder space for different classes of stock. Automated drinkers may occasionally malfunction and the lack of water may go unnoticed by the stockworker for some time. It is therefore important for birds kept within cages to have access to at least two separate drinkers.

The methods of rearing parents of meat-line domestic fowl give other welfare criticisms. The feed intakes of these flocks are restricted to approximately half the amount they would eat voluntarily. The birds are chronically hungry from the first few weeks of life until they reach sexual maturity. They often have abnormal forms of behaviour. However, flocks of parent breeders that have been restricted during rearing have much higher egg outputs and lower mortalities during their laying period than flocks allowed feed ad libitum during rearing. The welfare criticisms of this method of poultry production are therefore difficult to resolve.

Freedom from Discomfort

There are several criteria for cage floors to ensure that cage-reared poultry maximize their potential for growth or egg output. Floors should have a mesh that does not cause injury, that does not damage or lead to contamination of eggs, that allows birds to walk freely and that allows excreta to fall through easily. A galvanized wire mesh is frequently used to meet these requirements.

Preference tests show that poultry dislike using wire mesh floors and prefer to stand on littered solid floors. Further, there is evidence that birds housed for long periods on wire floors develop many foot defects, often claw fold damage on the front toes. The defects do not affect the mobility of the birds, but they do indicate a chronic discomfort from standing on this floor surface.

Foot health can be improved by ensuring there is a smooth galvanizing coat on the wire and by decreasing mesh size and increasing mesh gauge. A mesh size that allows the bird to stand on at least three toes of each foot is a suitable, practical compromise that still allows excreta to fall easily through the floor.

Sloping wire mesh floors cause additional discomfort to birds. Steeply sloping floors tend to cause an increased incidence of inflamed and swollen claw skin fold on the front toes of the birds. The slope of a wire mesh floor for caged laying birds is critical in ensuring the collection of uncontaminated eggs. Eggs laid in cages with a shallow floor slope do not roll out of the cage quickly and may remain close to the birds for some hours. The birds may peck at or stand on the eggs or excreta may land on top of the eggs. A slope of 8° is a compromise that allows good egg movement but reduces foot discomfort in the birds.

Freedom from Pain, Injury or Disease

Birds kept in groups may sometimes develop an unacceptable level of behavioural problems, often called vices. Feather pecking can cause high levels of feather loss in a flock. Feather pecking or vent pecking may develop into situations where birds cause serious physical injury to each other. Outbreaks of cannibalism can lead to the death of a large proportion of the birds within a flock. Cannibalism is a major concern of both stockworkers and welfare groups.

Beak trimming can be used to reduce the incidence of these vices. Beak trimming involves removing the final quarter to one-third of the upper beak of birds usually by a mechanically operated blade that then cauterizes the cut. Although the trimmed beak grows back to a normal length, nerves and sense receptors in the beak do not penetrate the scar tissue at the end.

Several factors may reduce the frequency of feather pecking and cannibalism. For example, a reduced light intensity or change of group size may be effective, but beak trimming gives the surest long term reduction in the incidence of these vices. There is acute pain for the bird when its beak is trimmed and there is evidence that the practice also causes long-term pain to the birds. It is, therefore, difficult to conclude whether the welfare of commercial poultry is improved when beak trimming is practised.

Freedom to Express Normal Behaviour

Some welfare groups believe that poultry cages provide a barren environment for birds and they decrease the birds' ability to perform natural biological functions. High stocking densities are commonly used in cage systems. Birds have a restricted ability to walk within a cage and there is a deprivation of wing flapping and flying. Many countries have laws or codes of welfare that state a minimum space allowance for caged poultry.

Simple cages also deprive laying hens of the opportunity to perform scratching and dust bathing behaviour. Furthermore, birds prefer to spend long periods of time perching and laying hens are strongly motivated to lay their eggs in secluded areas. It is difficult to provide practical solutions to allow caged laying hens to express these behaviour patterns. A cage design called 'a getaway cage' has been proposed that includes nest boxes, perches and a small sand tray for scratching. The getaway cage has not yet been successfully used in commercial egg production systems. Commercial egg production systems that allow nesting and scratching behaviour tend to be non-cage systems. There are many of these systems, ranging from free-range systems to colony systems within controlled environment housing (Fig. 9.4). All such systems generally use lower stocking densities when compared to cage systems and the costs of egg production are higher.

Freedom from Fear and Distress

Fear is a reaction to a perceived danger. Many fear responses occur during the processes of catching and loading at the poultry unit and the transport to and unloading at the processing plant. Large intensive commercial poultry units mostly keep their birds in a relatively quiet and constant environment. There is large distress when these birds are caught before transportation. Regular gentle handling of poultry reduces their fear of humans but this is not a practical possibility in large poultry units.

Transportation of poultry is the most important factor in increasing fearfulness in poultry that are being moved to processing plants. Hunger, thirst and fatigue are increased over long transportation times and there is an increased possibility of heat stress. Long transportation times increase the intensity of the fear response and some countries have introduced laws that control the length of time that poultry can be transported.

The information given in this section on welfare is not able to describe fully the scope and complexity of the subject. Current research on poultry behaviour and welfare is rapidly changing our understanding of this area of poultry science.

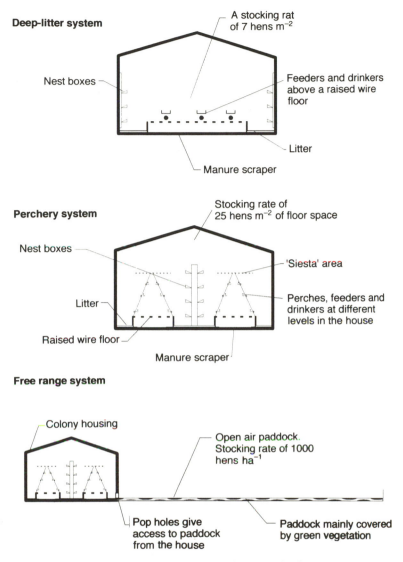

Deep-litter system

A stocking rat of 7 hens m^{-2}

Nest boxes

Feeders and drinkers above a raised wire floor

Litter

Manure scraper

Perchery system

Stocking rate of 25 hens m^{-2} of floor space

Nest boxes

'Siesta' area

Litter

Perches, feeders and drinkers at different levels in the house

Raised wire floor

Manure scraper

Free range system

Colony housing

Open air paddock. Stocking rate of 1000 hens ha^{-1}

Pop holes give access to paddock from the house

Paddock mainly covered by green vegetation

Fig. 9.4. Three colony systems for egg laying domestic fowl.

FURTHER READING

Housing and Environmental Control

Appleby, M.C., Hughes, B.O. and Elson, H.A. (1992) *Poultry Production Systems. Behaviour, Management and Welfare.* CAB International, Wallingford, UK.

Charles, D.R. (1989) Environmental responses of growing turkeys. In: Nixey, C. and Grey, T.C. (eds) *Recent Advances in Turkey Science*. Poultry Science Symposium No. 21. Butterworths, London.

Maton, A., Daelemans, J. and Lambrecht, J. (1985) *Housing of Animals*. Elsevier, Amsterdam.

Wathes, C.M. and Charles, D.R. (eds) (1994) *Livestock Housing*. CAB International, Wallingford, UK.

Welfare

Compassion in World Farming (1993) *The Welfare Argument*. Compassion in World Farming Trust, Petersfield, UK.

Faure, J.M. and Mills, A.D. (eds) (1989) *Proceedings of the 3rd European Symposium on Poultry Welfare, Tours, France*. World's Poultry Science Association (French Branch).

Mench, J.A. (1992) The welfare of poultry in modern production systems. *Poultry Science Reviews* 4, 107–128.

Savory, C.J. and Hughes, B.O. (eds) (1993) Proceedings of the 4th European Symposium on Poultry Welfare. UFAW, Potters Bar, UK.

UFAW (1988) *Management and Welfare of Farm Animals*. 3rd edn, Ballière Tindall, London.

Index